Concept and Expression of Time : Cultural Variations and Impact on Knowledge Organization

CONCEPT AND EXPRESSION OF TIME : Cultural Variations and Impact on Knowledge Organization

A. Neelameghan
and
G.J. Narayana

Published By

Ess Ess Publications
New Delhi
for

Sarada Ranganathan Endowment for Library Science
Bangalore

CONCEPT AND EXPRESSION OF TIME :
Cultural Variations and Impact on Knowledge Organisation

S.R. Ranganathan's
Five Laws of Library Science

Books are for Use
Every Reader his/her Book
Every Book its Reader
Save the Time of the Reader
Library is a Growing Organism

ISBN: 978-81-7000-697-8

₹ 650/-

First Published 2013

Published by:
Ess Ess Publications
4837/24, Ansari Road,
Darya Ganj,
New Delhi-110 002.
INDIA
Phones: 23260807, 41563444
Fax: 41563334
E-mail: info@essessreference.com
www.essessreference.com

for

Sarada Ranganathan Endowment for Library Science
702, 'Upstairs', 42nd Cross, III Block, Rajajinagar,
Bangalore-560 010
E-mail: srels@dataone.in Phone: 080-23305109

Printed and bound in India at Salasar Imaging Systems.

Contents

Acknowledgements		(*vii*)
SRELS - *K.N. Prasad*		(*ix*)
1.	Introduction and Conspectus	1
2.	Knowledge, Information, Knowledge Society	16
3.	Socio-cultural Evolution and Information	32
4.	Socio-cultural differences	56
5.	Time in Life Course	74
6.	Time and the Arts and Humanities (with special reference To Indic cultures)	91
7.	Time in Information Science	112
8.	Time in Historical and Some Evidence-based Studies	134
9.	Ontology and Representation of Time in Knowledge Organization Tools Used in Information Systems	145
Epilogue		164
References		166
Index		182

Acknowledgements

We express our gratitude to the Sarada Ranganathan Endowment for Library Science, Bangalore, for accepting this book as a publication of the Endowment.

We are grateful to the Ranganathan Centre for Information Studies, Chennai and Bangalore, for permitting the use of the papers on the subject published in several issues of the quarterly Information Studies.

We thank Smt. S.K. Lalitha of the Sarada Ranganathan Endowment for Library Science, for her assistance in checking the references and for preparing the mss for printing.

Sarada Ranganathan Endowment for Library Science (SRELS)

The Sarada Ranganathan Endowment for Library Science (SRELS), founded by Dr. S.R. Ranganathan in 1961, is registered as a Charitable Endowment with the Government of India (1963) and with the Government of Karnataka (2006).

The aims of the Endowment briefly are:

- To improve the library and information services in India
- To train library and information service personnel
- To promote the provision of efficient library and information service
- To apply research results to library and information studies

The objectives of the Endowment are:

- Furthering the cause of library and information service
- Organising Seminars and Workshops on recent developments in library and information science and the related fields.
- Organising periodically courses of lectures on recent

developments in library and information science and related fields by national and international experts

- Supporting the publication of these lectures and of other nascent ideas in the field of library and information science.

The Endowment is furthering these objectives by publishing a periodical entitled SRELS Journal of Information Management (formerly, Library Science with a slant to Documentation (from 1964 to 1988), and Library Science with slant to Documentation and Information Studies (from 1988 to 1999). Workshops, Lecture presentations and Seminars are also organized periodically.

There are three series of lectures organized under the auspices of the Endowment:

1. Sarada Ranganathan Endowment Lectures. (1965-);
2. Dr. S.R. Ranganathan Memorial Lectures. (1973-); and
3. Curzon-Seshachalam Endowment Lectures. (2001-).

K.N. Prasad
Executive Officer
SRELS

Bangalore

1

Introduction and Conspectus

In this book we discuss the differences in the concept, perception, experience and representation of 'Time' in different cultures and the impact of such differences in cultural frames on knowledge organization and the design and use of knowledge organization tools. It will be appropriate to begin this Introduction by quoting from Swami Atmarupananda's 'Cultural presuppositions as determinants in experience'[1]

> *"We inherit certain basic, unquestioned beliefs from our culture about the nature of ourselves and our world of experience, which in turn determine to a significant degree what and how we experience. This influences all levels of our experience, from the experience of self, time, and space, to our experience of bodily movement.*
>
> *"Cultural presuppositions influence the process of knowing. It is in and through our presuppositions that*

1 Atmarupananda, Swami (1995). Cultural presuppositions as determinants in experience: A comparison of some basic Indian and Western concepts, p. 20-42. IN: Concepts of knowledge: East and West. Papers from a Seminar held from 4-10 January, 1995, at the Ramakrishna Mission Institute of Culture. Kolkatta: The Ramakrishna Mission Institute of Culture; 2000. ISSN 81-87332-01-8

we experience or 'know' the world around us as well as the inner world or self.

"There are unseen influences on the process of knowing which influence the resulting knowledge as well. Presuppositions are one such significant influence.

"These days we speak in terms of the 'global village' and 'multiculturalism'. As a result of this new planetary awareness people are increasingly sensitive to the fact that 'different' does not mean 'wrong' or 'inferior.'

A "central experiential foundation is, when comparing different cultures, culturally based differences are found in the sense of time, space, self, relationship, and other broad categories of experience.

"All our perceptions, thoughts and actions take place in time, and so our concept of time obviously has a profound influence over us. We often hear about Western, linear time as opposed to Eastern, cyclical time. Linear time has an absolute beginning, a progressive development and an absolute end. Cyclical time sees universe after universe arising, developing, decaying, and being destroyed, without beginning and without end. The difference is real and significant. It is not that we choose to have a linear concept of time from a list of options. We grow up with a particular concept of time which we have unconsciously imbibed from our culture. It is just there, determining our mental and physical experience without our knowledge and without our conscious consent. Our life proceeds within this culturally defined sense of time, and we are quite unconscious of its nature unless it is brought to our attention.

"Greco-Roman time is generally considered to have been cyclic, not linear, but that has had little or no influence over the present sense of time in Western civilization. The Judeo-Christian linear concept of time, however, has

had a profound influence. Judeo-Christian time was not always exclusively linear. Origen, a great Christian thinker of the second century, believed that the present universe is only one of a succession of universes, and such views were held by some Jewish thinkers also. But from an early time in Christian history, linear time – with its belief in an absolute beginning and absolute end and progress in-between – became the accepted view. Linear time became the accepted view in Judaism, as well. And certainly the account of creation in Genesis and the rest of the Biblical revelation lends itself most easily to the idea of linear time.

"Not only is modern Western religious thought founded on this idea of linear time, but so is Darwinian evolution and its relatives, so in the Marxist view of dialectical materialism, so is the promise of technology on which most of the modern scientific endeavour has been based...... Social evolutionists similarly see society evolving, with a gradual reduction of evil and increase of good. Marxists see human history as a linear process leading through various evolutionary stages until the fulfillment, wherein the state withers away, being no longer needed. . Modern technology has driven the development of modern science largely through the premise of greater and greater control of the natural world, which will provide greater and greater good for humanity (good being defined here largely as comfort and security)

"This view of time is contrasted with the Hindu view, where time is cyclical. In cyclical time there is no absolute beginning or absolute end. History is not tending toward a final, glorious end. Every creation is followed by destruction which is followed by creation. Summer is followed by winter which is followed by summer. The sun rises and sets only to rise and set again and again.

> *Birth is followed by death which is followed by birth ... If there is evolution, there is involution. .... As for social evolution where evil is gradually conquered and only good remains, that is impossible according to the Hindu view. Good cannot exist without evil ... There are periods of increase and periods of decrease, golden periods and dark ages, but there is an overall balance of good and evil through time.*
>
> *"These differing views of time have profound effects on our inner and outer lives".*

A Unesco Expert Meeting on 'Time and the Sciences: Impact of Scientific Expressions of Time on Differing Cultures," London, 1976, resulted in the publication "Time and the Sciences: At the Crossroads of Cultures" edited by Frank Greenaway and Josefina Mena Abraham [Unesco, 1979] . The contributed papers reviewed and discussed the concept and expressions of time in different domains - the physical sciences, information science, the humanities, arts, architecture, philosophies, religions, cultures, etc. A. Neelameghan participated in the meeting and contributed the chapter Expressions of Time in information science and their implications. G.J.Narayana while working on his book "Knowledge and Information: Perspectives and Prospects" (2010) came across the above paper and thought that it was worth republishing it in a journal. Following discussions it was considered better to revise the text and add it as part of a more comprehensive treatment of the subject 'Concept of time in different cultures and the impact of the differences in the representation of time in knowledge organization and knowledge organization tool.' Hence the present book.

CONSPECTUS

The subject of this book is multi-focal - temporality, social and cultural knowledge, and their impact on the expressions of time in different domains including information science, and systems and services. All these facets are intertwined because they are the characteristics of one common entity, Time, and they influence each other. Information systems hold mainly embodied knowledge and their surrogates in document, non-document, and /or digital form. A variety of means and methods and tools are used for organizing, searching and retrieving the embodied knowledge. We therefore begin with brief notes on knowledge, and modes of valid knowledge. The concept of time is also elaborated in the first chapter. Certain expressions of time in information studies not influenced by cultural differences are discussed as also how these differences impact expressions of time in information systems is also discussed.

Following an introduction and conspectus or summary of the chapters in chapter 1. **Chapter 2** gives working definition and scope of selected terms. Discusses briefly the modes of valid knowledge – through the senses, intuition and extrasensory perception – and comparison of the knowledge acquired by these modes. Different categories of knowledge, social justification of large projects and the knowledge and the vast amounts of information thus generated and the role of the Internet in this context, are briefly dealt with. Some world views, such as physical sciences' view, cybernetics-centered new cosmology, and some of the approaches / perspectives and theories about the emerging society - information society and knowledge society are briefly discussed.

Chapter 3 deals with the socio-cultural evolution of human species. Evolution of higher forms of life gradually developing out of lower forms is a time-related process. Survival of species and how the living process changes with time are key areas of study in evolution. In human beings the evolutionary changes may relate to and affect cognition and related processes. The chapter also discusses survival of biological species including human beings, apart from genetic imposition, is dependent on their adjusting to the changes in the environment. Organization of human beings into groups based on affiliations such as kinship, geographic affinity, language, faith systems, common heritage, and the like is the basis of evolution and organization of human societies. A significant factor is that humans are endowed with a higher level of consciousness and thinking ability compared with other species. The ability to communicate with fellow members of the species through the language of speech, signs and symbols distinguishes humans from animals. At the same time, these unique features, genetic and mental attributes of individuals making up the society, and the socially constructed factors which include items like need for a vocation to satisfy one's needs and wants such as, food, clothing and shelter, means of optimizing production, finding new things and processes to economize and making life less stressful and enriched, besides the needs of mental solace, cultural development, social cohesion and harmony and spirituality. It is social organization and reorganization taking place on a continuous basis that combines and integrates the effects of these different factors over a period of time that shapes the features of a community or society. Information and communication are the twin critical elements in organization that achieve coalescence through exchange and sharing of thoughts and experiences among individual members and also institutions which help build the structure of a society. Enormous increase in communications involving a multitude of agents over a wide range of activities is a feature of advancing societies and also a symptom of increasing complexity. Communication takes place within a time-frame, which keeps changing.

The discussion also covers the growing complexity of social organizations and the related chaos theory viewed as a time-related factor. Ensuring continuity in society over time, time-binding and survival in time are among the social functions of communication. Also briefly discussed include: societal homoeostasis, time as a standard of comparison of features of societies (ancient, medieval, modern, post-modern, etc.) In the ICT-centred 'modern' society, the world is perceived to be shrinking in time and space (distance); and enormous volumes of information produced impact in different ways people's time and management of time.

Chapter 4 is about cultural differences: Concept of Time is almost universal in all cultures, but there are significant differences among different societies / cultures in the manner in which Time is measured / calculated, expressed and used in daily life. Cultural difference may arise from different faiths and religious beliefs, philosophies, different occupations, historical reasons, migration and immigration of people, etc. Why do some people seem to live their lives by the clock while others have a more laid-back approach to time? This may have to do with the values and expectations of the particular cultural group they relate to, grow up in and are conditioned by. In this regard there are differences between various "ethnic" groups, social groups, regional groups and even clans and families. So differences will be discernible even between different family groups or lineages in what might otherwise be considered as one ethnic or social group.

Some families or ethnic groups value abstract structures and schedules more than others. Some societies are highly structured. How things are done and how long it is expected for something to get done depends on the overall context of the event, the group expectations and how much value is put on time efficiency. All cultures have some concept of efficiency, but getting things done quickly is a value only in some mechanically-oriented or technological societies, and task-oriented or production-oriented situations.

Urban settings in every culture tend to be more "time-conscious" in regard to artificial and mechanical "clock time" than rural societies ("traditional" societies). The latter tend to be more relationship and community oriented, and values related to interpersonal relationships or group identity are given greater value than tasks or production, or related technological or abstract demands.

In task-oriented linear-thinking societies time is a commodity, whereas in traditional and more rural societies, which are generally oral-relational in character, time is a function of one's relationships or the cycles of life. Select examples of the concept of Time and expression of 'past', 'present', and 'future', new year, etc. in different cultures are briefly described in this chapter, with emphasis on Indic cultures.

Chapter 5 deals with Time and Life course: Temporality is an inherent element in life course. Life course is shaped by how social processes, such as, the family, education, and employment, health and migration domains are structured over the individual's life span Life events such as marriage, birth of a child, education milestones, career entry and exit, retirement, are important temporal events in one's life course. These events can change the direction of life course.

Life course forms the core .in a temporal perspective of social and cultural knowledge. This chapter is largely based on Melinda Mills exhaustive review of life course research. Taking an ontological approach, Mills reviews the temporal shaping of life course from a philosophical standpoint and from the epistemological standpoint. Data, methods and methodology are the epistemological instruments linking empirical evidence to the social processes that make up life course.

Life course perspective elaborates the importance of time, context, process and meaning in human development and family life, and focuses on the relationship between individuals and the historical, socio-economic, and cultural context in which these individuals live(d), how the past and present contexts and events impact on individual's behaviour and may provide clues on the future trends. Thus, temporality, the concept of time is inherent in all aspects of life course. Social processes including family, education, employment, health, migration, etc. are affected by temporality

Time dimension of social and cultural life is primarily made up of natural and cosmic phenomena, and cultural, social and institutional forms which together shape human life and society. Nature imposed effects determine physiological development of human body, such as time related aging process, menstrual cycle in women, and cyclic operation of seasons dictate agricultural practices, and the like. More and more institutionally framed calendar time determining our social, cultural and vocation practices impact on life course. Technological and medical advances, innovations and devices, such as air conditioning, contraceptives, and surrogate motherhood are reshaping cultural, social and institutional constructions in some countries. This has resulted in increased colonization of time in life course by neutralizing the effects of natural and cosmic phenomena

Briefly discussed also are different forms of temporality; natural and cosmic time, biological and psychological effects of time on human development; historical, cultural and social time, institutional time, harvest time, family time, time in business cultures, cultures without concept of time, etc. Allocation of time, monochromic and polychromic time, gender-oriented calendars are also described. Examples of time-related proverbs and sayings from different cultures are given

Chapter 6: Time and the Arts and Humanities with particular reference to Indic culture is the focus of this chapter. A unity of concepts and terminology discernible in different art forms of Asia –music, dance, sculpture, architecture - which parallels the concept of seminal mnemonics postulated by S.R. Rangnanathan. Indian classical music and dance forms have connections / roots in folk arts. Elements and aspects of the Indian cosmology, including expressions of Time, appears to have been absorbed into the Indonesian Balinese dances, and the architecture of Cambodia's Angkor Wat temples, as in Indian classical dances and temple architecture. Indian folk arts are briefly described. The Time- theory basis of raga classification of South Indian classical music is discussed. Indian classical dance is concerned with timing and pose; it is differentiated from Western classical dance which is more related to / concerned with, movement in space, Brief notes are given about time in relationship to psychic abilities, the time-related styles in literature, and time element in the classification of Classics in S.R. Ranganathan's Colon Classification

Chapter 7: This chapter on Time in Information Science begins with the notions of natural information and encoded information. Natural information is the pattern of organization of matter and energy in the material world. Natural information gets represented through the process of encoding and embedding. The encoded information is based on patterns of organization derived from the use of symbols, language, and signals and what we encounter in our social and perceptual world, Embedded information is the corporeal expression or manifestation of previously encoded information. Information appears as the external manifestation of knowledge in which form it is named 'explicit' or embodied knowledge. Hence, the study of time-space effects on knowledge is also relevant to information science.

The definition of information science adopted by *Information Science Abstracts* (ISA) of USA is "an interdisciplinary field concerned with the theoretical and practical concepts as well as the technologies, laws, and industry dealing with knowledge transfer and the sources, generation, organization, representation, processing, distribution, communication and uses of information, and communications among users and their behaviour as they seek to satisfy their information needs" There is a major shift of emphasis in this definition from library science / documentation bias to that of the Internet industry dominated world. The shift of emphasis is accentuated by the dynamic changing situation.

A definition of information science to be valid today has to accept and reflect upon the nature, form and complexities of information and information systems of the digital era. This change is seen as a natural consequence of the application of Ranganathan's fourth and fifth laws of library science. Compliance with these laws is directly related to economizing of time by the library / information user and of the information providing staff and to the effects of change, growth (physical as well new resources, techniques etc., replacing outdated ones). Change occurs in a time / space matrix, and has a reciprocal relationship with knowledge.

The changes over time in the production and quantification of information are discussed. The time-related attributes of time discussed include compressibility and expandability; role of communication in information flow; bibliometrics and informetrics; patterns of growth of information over time, the logistic pattern and exponential pattern; application of Boltzman equation model and heat conduction model to diffusion of information; sociological implications of time-dependent models; and acceleration of pace of information growth in relation to its obsolescence rates. Time in collaborative research and business data growth, statistical pattern in information phenomena such as document life parameters are considered. Sociological and behavioural aspects of research communication are also noted.

Chapter 8 : Time in Historical and Some Evidence-based Studies: This chapter briefly overviews a few subjects in which 'Time' is an important element. These subjects include History variously defined as (1) the aggregate of past events; the continuum of events occurring in succession leading from the past to the present and even into the future; all of human history; (2) records such as documents of different genres, providing permanent evidence of or information about past events in the form of written record - memorabilia, file, data file, account, chronicle, story; a set of related records, either written or in digital form kept together; includes working papers or records of activities involved in carrying out a project, etc. Viewed from these perspectives every subject, macro or micro, has a historical aspect.

Archaeology "is the study of human activity in the past, primarily through the recovery and analysis of the material culture and environmental data that humans have left behind, which includes artifacts, architecture, biofacts and cultural landscapes (that is the archaeological record). Archaeology uses a wide range of different procedures; it has a scientific component and a humanistic component. In some countries, such as, the USA, it is considered as a branch of anthropology, but in Europe it is considered as a separate discipline. (Wikipedia)

Archaeology studies human history from the development of the first stone tools some 3.4 million years ago in eastern Africa and continued to the present. Archaeological studies are important "for learning about prehistoric societies, when there were no written records for historians to study, making up over 99% of total human history, from the Palaeolithic era until the advent of literacy in any given society. The goals of archaeology cover a wide spectrum ranging "from studying human evolution to cultural evolution and understanding culture history." Archaeology also sheds light on many of humanity's technological advances at different periods of its development, e.g. the making and use of fire, the development of stone tools, the discovery of metallurgy, the beginnings of religion and agriculture. Archaeological studies are also extended to other planets.

Fossils : Fossils "are the preserved remnants or traces of animals (also known as zoolites), plants, and other organisms from the remote past times.

The study of fossils across geological time, how they were formed, and the evolutionary relationships between taxa (phylogeny) are some of the important functions of paleontology. Such a preserved specimen is called a "fossil" if it is older than some minimum age, usually the arbitrary date of 10,000 years ago. Hence, fossils range in age from the youngest at the start of the Holocene Epoch to the oldest from the Archaean Eon, up to 3.4 billion years old. The observations that certain fossils were associated with certain rock strata led early geologists to recognize a geological timescale in the 19th century. The development of radiometric dating techniques in the early 20th century enabled geologists to determine the numerical or "absolute" age of the various strata and thereby the embedded fossils. Estimating the dates of fossils is called geochronology.

Paleontology's goal is to map out how living things have changed over time.

Astronomy, one of the natural sciences, deals with the physics, chemistry and evolution of celestial objects, such as moons, planets, stars, nebulae, and galaxies. Also included are studies on phenomena that originate outside the atmosphere of Earth - supernovae, gamma ray bursts, and cosmic background radiation. Cosmology, a related subject, covers studies on the universe as a whole.

Astronomy is one of the oldest sciences. Prehistoric cultures have left behind astronomical artifacts such as the Egyptian monuments and Nubian monuments. Early civilizations such as the Babylonians, Greeks, Chinese, Indians, Iranians and Maya had carried out methodical observations of the night sky. With the invention of the telescope astronomy began developing into a modern science.

Specific subfields include: Solar astronomy, Planetary science,

Stellar astronomy, Galactic astronomy, Extragalactic astronomy, and Cosmology.

Observational astronomy includes: Radio astronomy, Infrared astronomy, Optical astronomy, Ultraviolet astronomy, X-ray astronomy, Gamma-ray astronomy, Fields not based on the electromagnetic spectrum, and Astrometry and celestial mechanics.

Archives: An archive is a collection of historical records, or a depository containing historical records and documents. Archives contain primary source documents that have accumulated over the course of an individual's or organization's lifetime, and are kept to get information about an individual or an organization.

In general, archives consist of records that have been selected for permanent or long-term preservation on grounds of their enduring cultural, historical, or evidentiary value. Archival records are normally unpublished and almost always unique, unlike books or magazines for which many identical copies exist.

The practice of keeping official documents is very old. Archeologists have discovered archives of hundreds (and sometime thousands) of clay tablets going back to the 3rd and 2nd millennia BC in sites like Ebla, Mari, Amarna, Hattusas, Ugarit, Pylos. These discoveries have been fundamental to know ancient alphabets, languages, literatures and politics.

Archives were well developed by the ancient Chinese, the ancient Greeks, and ancient Romans (who called them Tabularia). However, much of these have been lost, since documents were written on organic materials like papyrus and paper. On the contrary, many archives established since the Middle Age by churches, kingdoms and cities survive and often have kept their official status uninterruptedly even now. They are the basic tools for historical research on these periods.

Medicine / Healthcare: Apart from 'historical' studies briefly described above, subjects such as Medicine (healthcare) and Law depend on records of time. For instance, in the inpatient's record among other data the date of birth, age, the date and time of admission, duration of treatment, prescribed drug dosage and timings, the date and time of surgery, duration of the surgical intervention, the date and time of discharge, the date and time of re-admissions, etc. and recording of time-related data, such as, heart-beat, and pulse rate are required for subsequent use / consultation, clinical discussions, etc. as necessary, including when forensic (medico-legal) questions arise. Such case records data are also used in research, data analysis, statistical studies, and training.

Law : The practice of law is largely evidence-based and case reports, which contain the year, date and time of occurrence of events / happenings are researched / investigated and recorded.

Chapter 9 deals with Ontology and Representation of Time in Knowledge Organization Tools used in information systems. Ontology has philosophical and epistemological dimensions. The philosophical standpoint addresses social and cultural knowledge from the standpoint of temporality, the epistemological standpoint describes representation and organizational aspects of cultural and social concepts for later retrieval, which falls under the domain of information science and information systems. In designing systems for representing, organizing, and retrieving information it is ontology taken in its epistemological sense that reveals the first realities that deal with the fundamental categories of object, state of affairs, part, whole, and so forth, as well as the relations between parts and the whole and their laws of dependence (Albartazzi, 1996). Here ontology is "an explicit specification of a conceptualization" (Gruber, 1993). It is how the term is used in artificial intelligence and knowledge representation.

In computer science and information science, ontology generally means a formal representation of a set of concepts within a domain and the relationships between those concepts. It is used to reason about the properties of that domain, and may be used to define the domain. In theory, ontology is a "formal, explicit specification of a shared conceptualization", that is, an ontology provides a shared vocabulary, which can be used to model a domain — the type of objects and/or concepts that exist, and their properties and relations.

Ontologies "are used in artificial intelligence, the Semantic Web, systems engineering, software engineering, biomedical informatics, library science, enterprise bookmarking, and information architecture as a form of knowledge representation about the world or some part of it."(Wikipedia). Ontology in Artificial Intelligence (AI) refers to an engineering artifact, constituted by a specific vocabulary used to describe a certain reality, plus a set of explicit assumptions regarding the intended meaning of the vocabulary words. This set of assumptions has usually the form of a first-order logical theory, where vocabulary words appear as unary or binary predicate names, respectively called concepts and relations. In the simplest case, ontology describes a hierarchy of concepts related by subsumption relationships; in more sophisticated cases, suitable axioms are added in order to express other relationships between concepts and to constrain their intended interpretation (Corazzon, 2010). The creation of domain ontologies is also fundamental to the definition and use of an enterprise architecture framework. (WordNet 2.1). From another point of view ontology is considered as an enabling technology for the Semantic Web. It assists in stating the connotation / meaning of the data that one may generate or use or share with other information seekers. Here ontology enables people to state what they mean by the terms used in data / information that they generate, share, or use. Folksonomies are an emergent phenomenon of the Social Web. They arise from data about how people associate terms with content that they generate, share, or use. Gruber adds that as the Semantic Web matures and the Social Web grows, there is greater value in applying Semantic Web technologies to the data of the Social Web.

Wennerberg and Schulz write that to overcome confusion that may be caused by socio-cultural differences in expressing time there is a need for ontology of Socio-cultural Time Expressions (SCTE) The purpose of such an ontology is "to explicitly and formally describe these context dependent temporal expressions. This is done by defining appropriate hierarchical categories to which the expressions will be assigned. The ultimate goal is to be able to develop context aware temporal applications for the Web such as calendars or appointment schedulers".

2

Knowledge, Information, Knowledge Society

Information systems hold mainly embodied knowledge and their surrogates in document, non-document, and/or digital form. A variety of means, methods and tools are used for organizing, searching and retrieving the embodied knowledge. We therefore begin with brief notes on knowledge, and modes of valid knowledge. We also elaborate on the concept 'knowledge society' in this chapter.

2.1. DEFINITIONS AND SCOPE

Idea : An idea is a product of thinking, reflecting, imagining, etc., formulated by the intellect by integrating with the aid of logic, selections from the apperception mass, and/or what is directly apprehended by intuition. The meaning of each of the terms 'concept', 'thought', 'notion', and 'impressions' is near-synonymous with that of idea. (Ranganathan, 1967, Chapter CR, Sec. 2)

Concept : A concept is any identifiable mental image. Concepts carry out two broad functions: categorization and communication. Category is the class of objects. Categorization is the process by which mental representations, i.e. concepts, determine whether a given entity is a member of a category. This allows us to understand more about a new entity so categorized by virtue of the knowledge we already know about the category. Addition of a new novel entity to a category is likely to enlarge or limit the boundaries of that concept. Communication role of concept is largely taken care of by language (Medin and Rips, 2005). A concept is expressed through language. Thus, a concept is conveyed by words (vocabulary) vested with meaning (semantics). The basis of any information system is the existence of concepts, which in turn make up the world of knowledge. Thus, to understand and express unambiguously the notion of time specific to a society and the framework of time/period in which a concept is embedded an awareness of the socio-cultural environment plus the temporal framework in which the concepts and the languages evolve is a prerequisite.

Knowledge : Knowledge is the totality of the ideas conserved by human beings at any point of time. In this sense, the term 'knowledge' is synonymous with the term 'universe of ideas'.(Ranganathan. 1967, Chapter CR, Sec. 21).

Information is the manifestation of knowledge outside the mind; also called explicit knowledge. In the context of 'time in information', the use of the term'concept' (near synonyms: thought, notion), prevalent in cognitive psychology, is preferred to the term idea, as the former is more explicit in explaining the role of time in information systems and process.

2.2. MODES OF VALID KNOWLEDGE

Normal modes of valid knowledge according to Vedanta philosophy include:

- Perception;
- Inference;
- Verbal testimony;
- Comparison;
- Postulation; and
- Non-perception / Non-apprehension

- Extrasensory perception

These six modes are propounded by the Advaita School of Vedanta. Most other schools of thought concur with this categorization, but some have fewer categories, that is, some categories mentioned above are subsumed under a 'broader' category. Each of these modes has a characteristic way of conveying knowledge.

Each method of knowledge has "its own way and sphere of operation without contradicting one another. Perception is the means of immediate cognition. It can be external and/or also internal – the experience of the physical objects and the experiences of the mental states. Inference, comparison, and postulation, which are based on sense-perception, are the means of non-perceptual or indirect cognition, known directly by its non-apprehension or non-perception. Sense-perception and the three means of knowledge dependent on it can impart only the knowledge of sensible facts, but not of the supra-sensuous, whereas verbal testimony can convey the knowledge of both – the sensible and the supra-sensible. (Satprakashananda, 1974; Harshananda, 2008).

2.2.1. Perception

The noun perception has five connotations:

- way of conceiving something;
- the process of perceiving;
- knowledge gained by perceiving;
- sensing, perception - becoming aware of something via the senses; and
- percept, the representation of what is perceived is a basic component in the formation of a concept.

Perception may be external or internal. Perception by any of the five sensory organs (natural and direct way) is external. Mental perception (e.g. of hunger, pain, pleasure) of knowledge (or ignorance) love, hate, etc., is internal. Man's capacity for sensory perception is similar to that of other living creatures. Thus, perception is the primary mode of direct or immediate knowledge of physical entities, facts and of the mental states. This knowledge is got through sense-object contact or interaction (Satprakashananda, 1974).

Sensory knowledge (*jnana*) is acquired through the five sensory organs. They are evolutes of the sattva-part of the five primordial elements or tanmatras according to Vedanta (Harshananda, v2; p.114)

2.2.2. Intuition

Intuition is also a mode of knowing. It is instinctive knowing about something or apprehension of the given, without the intervention of rational processes that is, intuition is trans-sensory and trans-intellection. Mohanty (2002) tabulates the features of intuitive knowledge and non-intuitive knowledge as below:

Intuitive Knowledge	Non-Intuitive Knowledge
1. Immediate awareness	1. Mediate knowledge
2. The object is given	2. The object is constructed
3. The knowledge is non- conceptual	3. The knowledge is conceptual
4. The knowledge has absolute certainty	4. The knowledge may have only relative certainty
5. The knowledge is concrete	5. The knowledge is abstract
6. The knowledge is of an unique	6. The knowledge is of the general Individual
7. The knowledge is knowledge by identity	7. The knowledge is knowledge by difference
8. The knowledge is disinterested	8. The knowledge is motivated
9. The knowledge is an ecstatic awareness	9. The knowledge is detached, cold and intellectual

Mohanty argues that the above differentiation is not absolute, some pairs of contrasts coincide while others do not. Nevertheless it is a useful comparison for further discussion on intuition of different types.

2.2.3. Knowledge by Revelation

Protagonists of 'Revelation' consider that knowledge is revelatory. Its function is to manifest the object by removing from it the veil of unknown-ness, without affecting it in any way. It opens to view more or less what is hidden. It presents, but does not represent.

2.2.4. Knowledge Obtained Through Other Modes

Inference, comparison, and postulation are based on sense-perception and is indirect or mediate.

Non-apprehension and verbal testimony as means of valid knowledge have special characteristics. Non-apprehension is considered to be the only means of cognition of non-existence. Verbal testimony "has the unique property to communicate the knowledge of the supra-sensible as well as the sensible".

Extrasensory perception (ESP) is ability / power to perceive things that are not present to the senses.

2.3. WESTERN PHILOSOPHY

In Western philosophy, the study of the nature of knowledge began with the early Greek civilization with contributions from Pythagoras, Socrates, Plato, Aristotle, Archimedes and other philosophers. On this foundation, epistemology- the study of the nature of knowledge - advanced further under the impact of Renaissance and Industrial Revolution. It is in this milieu that such schools of thought as empiricism, materialism and rationalism got firmly established. It is around these three fundamental concepts that the modern epistemology in the West has developed branching out into new lines which include common sense realism, positivism, logical positivism, sense-data theory, phenomenology, realism, and social epistemology. According to classical epistemology knowledge is "justified true belief".

2.4. CHANGING MIX OF INFORMATION AND ITS SOCIAL IMPACT

The stock of information available in the world may be categorized into different types. Such a categorization helps in organizing information to serve a given purpose. It also shows the composition of the stock and its mix. Any change in the mix also serves as an index pointing out the changes taking place in the given social pattern over a period of time. The contents of this can be categorized in different ways. New characteristics of division may be introduced as necessary.

Information Categories : In earlier times by the term "information" mostly meant structured assemblies of concepts showing certain degree of homogeneity, exception being data type information. Such structured information implied value of some permanence and also formed the basic resource leading to the creation of knowledge. Besides structured information the active world also holds and uses enormous quantity of unstructured information or data. In this context it is helpful to examine Macklup's idea of classifying knowledge into (Machlup, 1980):

1. Practical knowledge. (Useful in work, decision-making and actions: Further divided into:
   - Professional knowledge. Political knowledge.
   - Business knowledge. Household knowledge.
   - Work-man's knowledge. Other practical knowledge.
2. Intellectual knowledge. (Satisfying intellectual curiosity as in liberal arts, education, and scientific research).
3. Small-talk and pastime knowledge.
4. Spiritual knowledge.
5. Unwanted knowledge. (This is outside one's interests, accidentally acquired, aimlessly retained)

In the emerging new order of society the demand for information pertaining to small-talk and pastime, commerce and consumer interests, and unwanted information constitutes an overwhelming portion of the totality of information floating in society. Demand and supply of sensual information outstrips the demand for intellectual information, i.e. information that goes to the creation of goods and services or enhance research and scholarship or helps managing. Most of the information products and services for entertainment, such as music, film, art, sports, and tourism are produced by industry and business, exception being the ones initiated by cultural and social organizations. Once these sectors are bestowed with the label of industry or business their primary goal shifts to profit making (maximum return in a minimum time period) and the emphasis will be on information dissemination (focused on target group(s) or on more wider potential customer groups) for product / service promotion and marketing.

(a) **Marketing :** It is consumer aimed. Marketing strategy is mostly created by advertising professions. In the consumer-aimed information (CAI), creation of brands and use of brand names form vital segments. Brands are built on the images of celebrities in various fields, but mostly from the world of glamour, sports, music and art, besides the reputation of the firm or that of the product / service as perceived by the customer. Development of marketing strategy has to keep pace with time.

(b) Preference for 'wants' satisfying information over 'needs' fulfilling information.

A predominant characteristic of the emerging new society is the higher value or utility placed on information satisfying senses or information having prestige value as seen by the masses, particularly among the younger generation. In this respect, we see a clear divide between the graying generation and the young. It is a known fact that in the hierarchy of needs among the poor and the poorest in developing countries, possession of TV and now, a cell phone takes priority over fulfillment of other needs like food, hygiene and health!

Social Justification of Large Projects : In the evolving information-centric world, it was stated earlier, that a large portion of the information held in society is of the ephemeral type - a category under Machlup's classification of knowledge. (Machlup, 1980). Ephemeral information has only momentary significance and marginal value. On the other hand, modern day scientific research is mostly collaborative in nature and often made up of large complex projects yielding huge data. Similarly 21st century global economy produces astronomical quantitaty of business data worldwide and most of it is converted into useful information. This has serious implications on time allocation, installation of requisite facilities, conversion of data into information, organization of data and information so produced, development of requisite skills to manage huge quantities of data, and their dissemination (see Chapter 5). These are long time projects involving international co-ordination, huge investments and operating on a continuous time basis.

2.5. EXPONENTIAL GROWTH OF INFORMATION AND INFORMATION SOURCES

Temporal aspect of information processing and providing access to information has two features, namely: a) shortening the time involved in processing and presenting information; and b) economizing the time of the information seeker. Both these factors are affected by the exponential growth of information (cf. S.R.Ranganathan's Laws of Library Science, especially the fourth law 'Save the time of the reader / of the library staff')

A person searching for relevant information in the ever expanding ocean of information is faced with serious problems in identifying information specific to his/her needs and in ensuring that no bit of relevant information is missed. Because of human nature to follow short cuts and easy ways, the searcher may be satisfied with whatever information is readily available, and thus burdened with both relevant and irrelevant texts. This in turn leads to feeding the information system with more and more unstructured texts not subjected to review process. It becomes a case of searching for a needle in a haystack. This may be avoided to some extent by acquiring the requisite information handling skills and tools.

Internet Media and Authorship : Every serious user of Internet and web sites who contributes to the content even in a small measure is a potential author/ source of information. The fact that one has created a web site means that he or she has some message to convey in a usable form, so that the text is more than an answer to a query or just presentation of data. It need not be scholarly. Internet and web sites being vast sources of information, structured or amorphous, refereed or un-reviewed, on all kinds of subjects, the intending author has a ready source of information to draw upon. Further, the use of e-mail and various social networks serve as ready sources of input information for writing and cross-checking.

2.6. WORLDVIEWS OF 21ST CENTURY SOCIETIES

To help summarize the state of society as it is evolving since late 20th century we have a few worldviews expounded by different pundits, namely a) physical view of the world, b) cybernetics centered society, c) information society; and d) knowledge society.

2.6.1. Information Bound Universe: Physical Sciences' View

Physical scientists view the world in terms of fundamental constituents of nature and universal laws governing the universe. The practical part of this worldview is accounted for by computation and computing, communication and electronic engineering, which explain the digital transmission and storage of information (including quantum information) phenomena in terms of electrical charges, bits, coding, and physical transmission; and limitations imposed by laws of physics on the speed of transmission of information, range of information processing available, and on storage capacity. (Landauer, 1991; Sigfried, 2000). Viewing from a higher metaphysical angle many physical scientists now believe that information is everywhere, it is distributed throughout the universe, and holding of information is an inherent characteristic of everything in the universe. Moreover, many scientists are now willing to concede that information belongs to the same fundamental category as that of time, space, matter, and energy. Rocks, atom bombs, black holes, and galaxies too register and process information. Every electron, proton and other elementary particles store bits of data; and every time two such particles interact those bits are transformed. Physical existence and information are inextricably linked. In other words, as proclaimed by Wheeler (1989), it is "It from bit" (Lloyd and Ng, 2004).

In this physical sciences' view, compressibility is a property of information that links closely information with physics, and indeed with all of science (Feynman, 1963). Applying brevity of expression and the use of command are the two most important means of imparting compressibility to information. Brevity retains cognitive links, which is an absolute requirement, and command works as a mechanical device.

2.6.2. Cybernetics-Centered New Cosmology

Taking cue from Whitehead's proposition that all organisms in nature exhibit a purpose or "subjective aim", which corresponds to the function of mind (Whitehead, 1934) Norbert Wiener built the theory of cybernetics. Cybernetics explains how systems maintain themselves over time through information feedback and processing of information reflecting the changing conditions in the environment over time. Information is defined by Wiener as the "name for the content of what is exchanged with the outer world as we adjust to it, and make our adjustment felt upon it. The process of receiving and of using information is the process of our adjusting to the contingencies of the outer environment, and of our living effectively within that environment" (Wiener, 1954, p.26-27). Cybernetics has given a new meaning to our perception of evolution and ordering of life. In his book The Biotech Century, Rifkin (1998) discusses the main features of the new biological world order and they are.

A living organism is no longer to be seen as a permanent form, but rather as a network of activity. Life is redefined as the philosophy of being, life, and mind become intricately bound to the notion of "processing" change. Evolution is seen as efficient processing of information. As an evolution process, negative feedback leads to stasis and positive feedback to transformation. Ability to process complex and increasing quantities of information by individuals and organizations has become the standard of measuring their success. Manipulation of genetic information to modify the future state of an organism and its environment represents the ultimate use of cybernetic principles and information processing skills. With every evolutionary advancement in the process of self organization of species, information feeds on further information to create more complex sets of relationships, which in turn increase the overall complexity of the system. The new matrix of evolution replaces the idea of life as machinery with the idea of life as information. By resolving structures into functions and reducing functions to information flows, the new cosmology all but eliminates the idea of species integrity and all living beings are drained of their substance and turned into genetic information. In the revised thinking about evolution everything is pure activity. Information has become the primary object of our attention and evolution of life. With information's new exalted status, even knowledge itself is loosing its meaning and increasingly being subsumed under the category of information. In the new paradigm, 'to be knowledgeable' and 'to be informed' have come to mean virtually the same thing, which means a revolution in the history of consciousness. Here, being informed is an ongoing process of anticipation and accommodation to changes going on in the environment; and knowledge is no more viewed as discovery of facts, but rather as an ongoing creative process.

Since knowledge is created so fast that today's awareness is getting outmoded by tomorrow. Thus, we have far less tolerance for timeless truths and ironclad laws of the earlier paradigm of science, which by their very nature, imposed boundaries and established limits. Boundary implies telling us what we can do and cannot do. Today, change itself is taken as the only truth. By reinterpreting nature as the evolution of information, new information cum change-centric cosmology has evolved.

Information Society : The predominant role played by information in a technological environment, supported by computer and communication technologies, in subordinating the importance of other factors of production in the economy led to the notion of information society also termed 'postmodern society' or 'postmodernism'. The term information society is primarily applied to the countries, which have progressed from the stage of industrial society in the late 20th century, and more particularly since 1990 (Hill, 2005). Original authors of information society are: Machlup (1979, 1980-84), Porat (1977), Bell (1976), Toffler (1970, 1980) and Naisbitt (1982).

There are different viewpoints on the characteristics and enabling forces which go to determine the contours and the image of an information society. These viewpoints are based on economic criteria, religious and cultural characteristics, socio-technical perspective, and value approach. According to Frank Webster an information society has five perspectives, namely: technological, occupational, spatial, cultural, and economic (Webster, 2002). These approaches are briefly described.

Economic Criteria : Central to this perspective is that by employing computer, Internet, web-technology and other electronic forms of communication in economic spheres, information energizes other factors of production to yield optimum results. As a result: a) the service sector overtakes manufacturing and other sectors of the economy and becomes the main single source of national income; b) information generation and supply becomes a major occupation and forms the main source of employment; c) information is valued above tangible goods, and d) one strives to make the maximum use of information (Hill, 2005). Objection to this exposition is that it gives only an apparent view. (See also Marschak,, 1968)

Religious and Cultural Viewpoint : Religious and cultural factors have a strong impact on the formation of one's attitude to concept of society. Collectively these attitudes may develop into an ideology or dogma which may promote or inhibit a society becoming a knowledge society.

Webster (2002) attributes the cause for culture-determined information society to the pervasive hold of media disseminated information, and the impact of anywhere, anytime personal and interpersonal informational intercourses on the society. Social intercourse now involves a greater degree of information content than ever before. It is information of all types bombarded from all sides that shapes fashion, individual life styles, and the social values. Such a society is an information-laden society, where information is received and exchanged instantaneously and its social and cultural adoption too is almost immediate. The conclusion is that experientially the idea of an information society is easily enough recognized, but a definition of the new society is less clear than any of the notions stated so far.

Socio-technical Theory : Webster identifies socio-technical perspective with the visible features of the information society created by the cable and satellite television, widespread use of computer (particularly personal computers) and communication devices, mobile phones and other hand-held devices, new office technology processing (particularly online services), and CD-ROMs. The socio-technical assessment of the present knowledge society shows that, unlike its predecessors, the present society is to an unprecedented extent the product of its own action, in which power may not be evenly distributed throughout the society. The balance between the power of nature and the power of society, i.e., the human capacity to control nature, has tilted enormously toward the latter (Stehr and Meja, 2005). However, Webster points out that technology is an asocial phenomena which cannot define a social world.

Value Approach : This approach relies on the impact of information and knowledge on the cultural values, faith systems, ethics and life style of people and its consequences for the individual and the society as a whole. The main criterion is the intercultural effects induced by new modes of communication systems and the capacity of globally accessible information to alter cultural and social values, and life styles of people. Value includes goodness value, ethical value, democratic values, and other higher level values. (*see also* Hirshleifer, 1971) (Pritchett, 2008)

Occupational Theory : According to Bell (1976), a knowledge society is manned and managed by knowledge workers. The people constituting the new work force bear the designation "symbolic analysts", "knowledge experts" or "informational labor". Their main job is to create and use knowledge / information in an endless cycle. This thesis is faulted on the ground that it is based on a count of raw numbers of "information workers".

Spatial Thesis : This theory is based on the overwhelming impact of information networks — a spatial or geography related phenomenon. In the information society homes, offices, businesses, recreational spots, and even religious and spiritual centres are all connected to one another by information networks. Castells (2000, p.3) observes that "our societies are increasingly structured around a bipolar opposition between the Net and the self". This has profound effects on one's organization of time and space. The logic in calling a society an information society merely because of the overwhelming presence of networks is questioned.

2.6.3. Knowledge Society

Knowledge society may be considered to be an evolving form of information society. If information society represents one face of information-knowledge duality, knowledge society represents the other side. According to existing perceptions, the main forces driving a knowledge society are: a) capitalism, b) globalization, c) liberalization of economy, d) wide spread use of information and communication phenomena, especially digital communication and networked information, e) social, cultural and religious attitude of a society, and f) stage and state of development of a society in terms of natural resources, intellectual resources, educational standing, and level of technology development. However, a knowledge society is to be determined by the criteria of its inclusiveness, that is, how far knowledge becomes an integral part of everyday life of all citizens, and in what ways knowledge is put to use by them to increase overall goodness in society; and not solely by attaining the state of a knowledge-based economy or by mere wide-spread use of information networks. Chevalier and Buckles (2008) infer in the context of "Social Analysis of Systems" that at present, the knowledge base of the knowledge society consists of knowledge resource generated, accredited or communicated predominantly in scientific, corporate and university-based settings that exclude and ignore many other segments of society. A true knowledge society has to engage all human beings without exception in the application and cogeneration of knowledge. Such a system draws on the information, imagination, skills and meaning, and reasoning of many people, seeing their different views and the methods they use as "living knowledge" that has a potential to advance the common good on a universal scale.

Normally most major social changes evolve slowly, at an even pace, and their impact is visible only after the completion of the transition, whereas in the knowledge society significant social, economic, and cultural changes that are becoming visible seem to indicate that they are of extraordinary present and future significance (Stehr and Meja, 2005). On the psycho-sociological front, society is subjected to more anxiety and increased sensitiveness to issues and happenings. This is because of increased diversity of functions, increased number of time bound activities to be completed, which means a pace of living dictated by tight time schedules. Even a small mishap or lapse occurring on an odd occasion is sufficient to cause a disproportionate amount of disappointment, and mental agony. Looking at it this way, there seems to be some truth in the saying "ignorance is bliss".

A temporal feature of knowledge society is that overall expenditure of energy, mental and electrical, per unit of society's time will be far greater than it has been in the earlier types of societies, although there may be saving of energy per unit time in many spheres of social and economic activities mainly due to the effects of technology application.

The idea of knowledge society has become popular with a section of the elite in some developing countries, such as India. If India is to become a knowledge society, it has to first tackle some of the basic issues facing the country. The issues relate to: a) changing the mindset of people towards access to information and its optimal utilization, which are linked with computer literacy and education; b) increasing the viability of rural India and urban poor, so that they become a part of the knowledge environment; c) changing the existing laissez-faire / negative work culture in the country; d) checking the mushrooming population growth; f) controlling corruption; and g) empowering woman. In this context, the recommendations of the National Knowledge Commission are very pertinent. This calls for the nation to produce the required political and moral leadership necessary to attain the state of knowledge society.

There are several contradictions in the current state of the knowledge society. To give one example is that in the knowledge society "the world opens up and creeds, styles and commodities mingle, yet the walls between incompatible convictions about what is sacred do not come tumbling down" (Stehr and Meja, 2005, p.15). Martin (1995, p.1) after studying the claims made by many that information society exists here and now says that, in the absence of credible evidence, the prevailing reality tends to indicate that "information society has remained little more than a catchphrase or a slogan, more a matter of rhetoric than of reality". This assessment also seems valid in the case of knowledge society as it exists today. But there is ample hope that true knowledge society will evolve sooner than later - a society guided by principles of harmony, sharing, equality, fellowship, and universal brotherhood.

3

Socio-Cultural Evolution and Information

3.1. SOCIO-CULTURAL EVOLUTION OF HUMAN SPECIES

According to the doctrine of evolution higher forms of life developed gradually out of lower forms. This evolution is thus a time-related process. The key areas of study in evolution include survival of species and how the living process changes with time. In human beings the evolutionary changes may relate to and affect cognition and related processes.

Survival of biological species is dependent on their adjusting to the changing environmental conditions, apart from genetic imposition. Human species is no exception to this. Organization of human beings into groups based on affiliations such as kinship, geographic and neighbourhood affinity, language, faith systems, common heritage, and the like is the basis of formation of human groupings, communities and societies. Humans are endowed with consciousness and thinking ability, which is not present in most other animal species to the same extent. The ability to communicate with members of the species (and of the group or community) through speech, signs and symbols distinguishes humans from other animals.

Over the centuries people developed a variety of natural languages and scripts that helped communication. At the same time, this unique feature, genetic and mental attributes of individuals making up the society, and the socially constructed factors which include such items as a vocation to satisfy one's needs and wants like food, clothing and shelter, means of optimizing production, finding new things and processes to economise on time and making less stressful and enriched, besides the needs of mental solace, cultural development, social cohesion and harmony and spirituality. This social organization and reorganization is a continuum combining and integrating the effects of these different factors over a period of time that shapes the features of a community or society. Information and communication are the twin critical elements in organization that achieves coalescense through exchange and sharing of thoughts and experiences among individual members and also institutions which help build the structure of a society. Fast buildup in communication capabilities involving a multitude of agents over a wide range of activities is a feature of advancing societies and also a symptom of increasing complexity. Communication takes place within a timeframe, which keeps changing. (*See also* Ranganathananda, 2008).

3.1.1. Complexity

Complexity is a time-related phenomenon that works in two ways: complexity as it appears in nature, and complexity in the social world. The former belongs to the domain of science, and the latter comes under the domains of social sciences and the humanities. Complexity as a scientific phenomenon is studied, among others, by chaos theory specialists.

3.1.1.1. Chaos Theory

Chaos theory or complexity theory, is a scientific discipline which studies nonlinear systems and explains the relationships between interacting elements in the system. The relationship is studied using models. An essential element in chaos theory is unpredictability. The cause of unpredictability is sensitivity to initial conditions. Very small changes in the starting conditions of a system set off dramatically different effects in the system. Edward N. Lorenz was the first to explain this phenomenon in his talk titled 'Predictability: Does the flap of a butterfly's wings in Brazil set off a tornado in Texas?' (McGill, 2007). Chaos theory finds application in a number of fields ranging from quantum mechanics and cosmology to economics and music. In biology chaos theory has helped in extending our understanding of "the new evolutionary processes leading to understanding of the genetic algorithm, artificial life simulations, learning process in systems including the brain, and studies of such previously considered unresearchable areas as consciousness and the mind.....Darwin's work also appears in direct conflict with Newton's because it changes our understanding of the nature of time, demonstrating that some time is not reversible" (http://library......chaos.html). Chaos theory is already impacting many critical aspects of our lives. Of particular interest to information/software field is fractal mathematics used to develop improved information compression, and encryption schemes needed for computer networking and telecommunications. Engineering applications range from factory scheduling to product design.

3.1.1.2. Socio-cultural Complexity

With passage of time, complexity increases in the human world due to dynamic and innovative nature of mind. Sociologists and anthropologists explain the cause and effects of social and cultural changes in a society called 'sociocultural evolution (ism)', which is made-up of two themes: social evolution, and cultural evolution. Its aim is to describe how cultures and societies have changed over time. Its theories typically provide models for understanding the relationship between technologies, social structures, the values of a society, and how and why they change with time; they vary as to the extent to which they prescribe specific mechanisms of variation and social change. A related concept of immediate relevance in interpreting and explaining the changes taking place in the present day world, and their cultural impact is 'sociocultural development'. Sociocultural development is a unified process of integration involving increases in the socio-cultural complexity that describes modern society (http:// wiki.sociocultural....). Both these processes are entwined with knowledge / information in a continuous cycle of change. The impact of these processes is examined in some detail from the point of view of information.

3.2. SOCIAL FUNCTION OF COMMUNICATION

3.2.1. Continuity in Society

A social function of communication is to ensure the continuity in society – in the present and with the past - through access to the experiences and ideas of the past, expressed in cultural symbols (e.g. language, icons, images, etc.) across space and through time. These are the 'time-binding' and 'space-binding' functions of social communication. In recent times, social networks, blogs, twitters, youTube etc. on the Internet have speeded up and widened the scope and reach of such communications. In the 1920s, Korzybski (1924) proposed a theory of human behaviour, based on binding ability, which provides an insight into the links and differentiation between plant, animal and man. In recent decades, studies in cybernetics, thermodynamics, ethology, neurobiology, etc. have provided supporting evidence and data for the theory.

3.2.2. Time Binding and Communcation

Man's time-binding ability arising from his use of language, number, gesture, picture and other symbolic forms, enables him to transcend the limitations of inherited characteristics and the seemingly insurmountable barriers of time and space. The insight, the wisdom, the culture and the arts, the ideals of the political, economic and social structures of the world, and the human values and experiences of past generations can be preserved, transmitted, used and evaluated by people both in the present and in the future thanks to the time-binding and culture-binding mechanism of symbolic communications. This implies the 'culture concept' expressed by anthropologists, linguists and sociologists. An animal lives directly in the 'present', without any apparent notion of the animals of its kind that have lived in the past before it, and of those that may come after it in the future. Man, on the other hand, because of his capacity for symbolic communication, has a sense of events in sequence - i.e. a sense of time. From this arises his motivation to transcend time, as evidenced by the great depth and breadth of his cultural borrowings from the past, using or practising them in the present, and passing them on to future generations. Man's superiority in the competition for self-perpetuation arises from his ability to use signs to indicate entities and to represent them. It has been pointed out that "the use of these substitute signs is the essence of symbolism and our power of reasoning; they can take the place of past experiences and feed the imagination by combining memories of things that might be in future experience; such an experience takes man from the here and now" (McGarry, 1975). Further, the ability to transfer information received by one sense mode to that of another - that is, cross-modal association - appears to be unique to man. It endows him with special creative ability to plan ahead and influence the future (Neelameghan, 1979).

3.2.2.1. Societal Cohesion and Communication

Human civilization and culture, organizations and institutions are held together over time in a network of mutual dependence and responsibility, and the sustaining transactions are carried on by myriads of messages communicated over this network. These messages form the basis as well as the regulating, organizing and time-binding influence in social evolution and progress. A failure in such communication could lead to misunderstanding, misinterpretation and confusion of the import of the messages (Neelameghan, 1979). In turn, these may cause alienation, aggression and detraction from human progress. Here communication connotes a deep, dynamic and complex series of processes operating together - perception, memory, and information analysis; and processing, symbolic transformation, delivery and feedback. The blending of the resources of cybernetics, information theory, systems approach, and semantics with those of science and technology has extended human foresight. Man, thus, appears to be moving toward the possibility of influencing the future and directing his own evolution.

3.2.3. Survival in Time and Cultural Co-Operation

Survival for man implies three facets, namely, survival in energy, survival in space, and survival in time. It involves physical inheritance and competition as well as cultural inheritance and competition, that is, competition in, with and for symbolic forms used in communication. Intra-species competition reduces the chances of survival of the species. Therefore, co-operation within species is essential to the physical survival and time-binding culture-transmission activity of man. Man is a link in the vast transmission chain of culture through time.

Knowledge and culture transmission and the perpetuation in time of society are secured not only through oral communication, but also through various social structures and forms, such as rites and rituals, sacraments, ceremonies, and institutions. Language is itself one such structure. The continuity between speech, writing, print, books and libraries defines language's time-binding institutional function (Evica, 1963).

3.3. SOCIETAL HOMEOSTASIS

3.3.1. Information and Entropy

Human society can be looked upon as a dynamic open system consisting of individuals, groups, organizations and institutions of various kinds in continual interaction and interchange of energy and messages (information). Here we shall primarily consider semantic interchange. Open systems have certain common characteristics, such as the following (Bertalanffy, 1968).

Input : Import of energy from the environment. Man, as a biological and social entity, seeks and inputs energy and the information and knowledge necessary to satisfy his biological, emotional, intellectual, economic, spiritual, and cultural needs.

Throughput : The input is processed by performing work and is converted into the output. Man himself is an information-processing system. He has also created various information-processing systems to help him

Output : It includes inference, data, ideas, description, information, knowledge, documents, and other information and knowledge sources.

Cyclic Character : Input, processing, output, feedback, control, adjustment and adaptation - each phase supports the next in a cyclical way. From this succession of events in time arises the interrelationship of the parts of the system and of their functions.

Negative Entropy : As per the second law of thermodynamics, the universe is running down. The energy loss or increase in entropy corresponds to the positive direction of time. Thus time's arrow, or probability vector, affects the physical, biological and communication patterns in the universe.

Some of Korzybski's time-binding agents, man in particular, constitute these local enclaves trying to maintain order. The commands through which man exercises control over his environment are a kind of information imparted to it. Increase in the supply of information increases negative entropy, helps to reduce uncertainty, imposes order, and prevents the system from tending towards chaos and disorganization.

3.3.2. Information Input, Negative Feedback, and Coding

Negative feedback is the information on what, if anything, is wrong with the system, and why it is necessary to make adjustments and take corrective action. Signals received may be expressed as codes. Man has evolved elaborate coding systems to facilitate communication through space and time.

3.3.3. Steady State and Dynamic Homoeostasis

The change in the attributes of the parts of a system over a time period t1 to t2 may not be drastic (steady state). On the other hand, information received may cause substantial impact, even cognitive dissonance, on the individual. Just as the human body is kept in dynamic equilibrium, that is, homoeostasis, through cell co-operation and communication, human societies depend upon cultural communication for their homoeostasis. Hence, co-operative behaviour patterns and the resulting meaningful and effective human relationships have a high survival value over time. These behavioural patterns and changes thereof are based on acts of symbolic communication.

3.3.4. Differentiation and Equifinality

Division of labour and specialization of functions and roles of the different parts of a society are necessitated by the growth in time of the size of the society and the complexity of its demands.

The possibility exists of attaining the set goals of the system, starting from a variety of initial conditions, by adopting suitable alternative means. The different phases of working of the system are time-dependent, and such processes and events as entropy and steady state are referred to or defined specifically in terms of time and information communication.

3.4. NEW SOCIAL ORDER

3.4.1. Time as a Standard of Comparison

Time or period is one of the means used by historians and even laymen in describing a society. Applying this standard a society is characterized as ancient, medieval, modern, postmodern, 21st century society and so on. According to Western thinkers, technologically and economically advanced societies in the West are said to be postmodern, i.e. the stage after industrial era. Indian society has reached modern stage, and is aiming to advance towards postmodernism / knowledge society, which represents a new social order. However, there are still communities around the world, though small, living untouched by modernism. Some of the main features of the new social order and the forces behind it need examining.

The societal homeostasis, and also the role of communication in the evolution of societies described earlier refer to the societies of the 1990s and earlier. Now there is a change of gear. Presently, the issue is not of survival in the biological sense (excepting a few concerns like, maintaining biodiversity, reducing greenhouse effect, and preventing nuclear holocaust), but unfolding of a new pattern of societal homeostasis, i.e. social evolution. In the new social order an individual is constantly confronted with the rapid changes caused by his unending interaction with the forces and elements in the environment, more particularly of his own making. Exposure to the enormous changes taking place in the man-made environment, in turn, modifies the cognitive behaviour of the individual. Once an individual steps out of his/her house or picks up or makes a telephone call or switches on TV or opens up the laptop his or her cognitive state is impinged upon as never before. The impingement generally results in a change in the mental state and behaviour. This is a continuous process. In the evolving society today, the nature, number, and the variety of impingements taking place is fast and immeasurable. The two revolutions that are causing and sustaining the new society are: a) the Internet (computer) and communication technology driven, and b) cybernetics centered. The new features, and forces released by these two revolutions, and their consequences on society are briefly discussed below, Neelameghan (1973) briefly described the situation and progress in developing countries.

3.4.2. Internet and Communication Technology-centered Society

The Internet and communication technology-centered view of society is shaped by the impact of computer and communication technology revolution. While the computer part of the revolution is accounted for by mainframe, tabletop, laptop and palmtop computers, the role of communication technology is represented by what is called mobile industry made-up of telecom operators, device manufacturers, platform service providers and content owners. If the constant theme of the industry is miniaturization of the gadgets, the technological objective is to blend the features of mobile and digital, which has enabled the introduction of tablet PC and smart phones in the market. New features are added to the existing set of gizmos almost every year. Network technology is extended from 2G spectrum to 3G and 4G. The agencies are the Internet, websites, social networks and a variety of channels dedicated to endless dispersal of news, views, knowledge, music, pictures and information relating to sports, recreation, entertainment, and others. The salient temporal features of the new social order are: time allocation, abridgement of time, and time tradeoff. The scope to exercise an individual's liberty and free will is all the time expanding. The new system is driven by information in all its forms and modes supported by cybernetics, bioengineering, use of feed-forward mechanism, and unbridled consumerism. This information-centric social evolution has many implications.

3.4.3. Extension of the Boundaries of Information World

Functioning of the universe, including life, and our understanding of it takes place at three levels – mega, micro and intermediate, with the last one corresponding to human perceptional range. In the early phase of human history our concept of reality, our understanding of world and life, was limited strictly to the intermediate level, that is, what our sense organs could mediate and the mind could think of. Advances in technology, in particular instrumentation technology since the eighteenth century showed that there is knowledge or reality beyond the intermediate level. The new vista opened up the universe of the mega and the micro. Stretching the universe in the mega and the micro directions meant extending the boundaries of the world of knowledge / information; and also discerning the variety, depth and complexity surrounding life and nature. Two of the earliest instruments invented to probe into the nature of the universe and life at the mega and the micro-range were telescope and microscope respectively. Development of various types of sophisticated optical, magnetic, acoustic and electronic instruments coupled with computer and communication capabilities have been responsible for most of our present-day knowledge, particularly relating to scientific, technical, medical, genetic, and space achievements. For example, spectrum analysis has enabled us to analyze the chemical and energy compositions of different stars, radio astronomy opened up the vista of distant galaxies and their stars, tele-communication systems have enabled communication with astronauts orbiting in space, cyclotron enabling splitting of the atom and thereby exposing the secret of nature at the tiniest level, and Magnetic Resonance Imaging (MRI) making it possible to map the active brain have extended the reach of human sense organs to collect data and information, beyond imagination. Instrumentation coupled with human drives extended the reach of our sense organs, opened up the mega and the micro universe, and enlarged the boundaries of knowledge / information world (Narayana, 2010). This led to questioning the validity of the then prevailing notions about life and nature at the turn of each major discovery / invention, which in turn impacted and continue to impact on our perception of reality and social values. Exploitation of the benefits accruing from the developments in science and technology in everyday life constitutes one of the cornerstones of the evolving social order. What is important to note is the speed with which it is taking place and its range of coverage. Hence, time has ceased to have its waiting effect as information can be spread to all corners of the globe almost instantaneously. At the same time, for individuals (and also institutions) time is becoming scarce as a day has only 24 hours, and brain has limitations to receive and process information.

3.4.4. Information Flood

Until three decades ago exposure to information was largely restricted to the four walls of a library, a school, a newspaper or a cinema hall, a radio or first generation TV, a family or group interaction. One had to access information (and also exchange) in a specified space or through a landline telephone or face-to-face meet all of which were subjected to time (and space) constraints. This has changed now. Today, the constraint is not one of providing access or bridging the time and space gap between the creator/server of information and its beneficiary. The issue is no more limited to access and flow of information, but adapting to the flood of information bombarded to each and sundry who has access to a computer, on a TV screen, and other media, coming free or on payment, or through a mobile phone and its updated versions constantly flooding the market. Literally, certain types of social and commercial information is bombarding the public whether one wants it or not by numerous information generators and promoters, which includes websites, twitters, social networks, TV channels and mobile phone. To protect individuals from being the target of all sorts of advertisements and phone messages aimed at promoting that product or this service, genuine as well as spurious, it has become necessary to enact laws to prevent unsolicited interruptions on one's private time and space. Flooding of information is not limited to sale of products and services, but now covers more and more issues of public interest and political agendas. Exposure to such a flooding of information has enormous impact on individuals causing behavioural change, affecting cultural values, changing social norms and promoting imitation proneness, and eroding traditional and existing social patterns.

3.4.5. Impact of New Media on the Public: Shrinking of the World in Time and Distance

Compared to the power and impact of the electronic media (such as the Internet and websites) information relayed through traditional media, such as, newspaper and printed book, the latter medium looks outmoded and lethargic. If search engines like Google and websites like Wikipedia help promote awareness on an unprecedented scale among peoples around the globe, news channels like Aljazeera can alter history in relatively a fraction of time it might have taken a decade or so ago. This comes from their ability to put enormous amount of information in the public domain, which is easily accessed and spread without much hindrance. These new media are not easy to control as they have worldwide spread, cutting across regions, religions, languages, and political borders. They have become the agents for carrying knowledge / information capable of provoking people, influence their views and behaviour, and cause social revolutions. Now there is no place to hide information from ordinary people, unless it is covered by intellectual property rights or kept under unbreakable secrecy. Another example is the WikiLeaks. All this goes to show how fast changes are taking place in the political, cultural, social and business arena and how the spread of information is overcoming the constraints of time, distance and national boundaries. Today information is accessible anywhere, any time by any one. Only limitations are: availability of telecommunication infrastructure, unfamiliarity with the operation of the system, and in many cases monetary constraints. On the positive side the spread of use of cell phones / mobile phones in the rural areas is enabling even rural folks easy and immediate access to information relating to product markets and prices, agriculture, health and sanitation, and such other matters.

The use of information power to influence people is nothing

new. From historic times spoken word and written letter have played an important part in bringing change in the lives of people, groups of people, and larger social entities. All revolutions, whether religious, political or social have depended on the word to spread the intended messages and induce individuals and masses to act on the messages. What is new in the emerging social order are the factors of time and place, tenor of delivering messages, and the type of content in the message. An example is the instance of the Rio Declaration regarding environmental issues. The Rio Declaration on Environment and Development, (Rio Declaration), was a short document produced at the 1992 United Nations "Conference on Environment and Development" (UNCED), the Earth Summit. The Rio Declaration consisted of 27 principles intended to guide future sustainable development around the world. It was perhaps the first time that non-governmental organizations (NGOs) could set up conference news reporting facilities at such a world summit. Even as the resolutions were being discussed and debated the reports could be sent around on the NGO networks. In a particular instance one of the developed countries did not agree to the wording of a resolution. However some days later the same country sent a 'revised' form of the resolution which it would accept. This modified text was immediately broadcast on the network. An environmental activist in India got the draft copy and found it was not acceptable to developing countries. The TWN (Third World Network) was alerted. The Kualalumpur HQ of TWN relayed the message to other heavyweight NGOs such as 'Green Peace' in Geneva and within hours an objection to the new text reached the Earth Summit HQ and the text was withdrawn soon after (Akhtar, 1993). This is an example of fast global distribution of messages enabled by developments in ICTs.

Implications of new media developments are:

1. Unlike in the past, a message / information is accessible to any number of people anywhere in the world including outer space simultaneously, thus nullifying the time divide and time delay.
2. Information in all forms and modes can now be uploaded and

downloaded via a desktop computer or a handheld device, and also relayed via smartphones. This provides the information seeker the choice to select one's own time and place in accessing uploaded information. Thus, time is seemingly brought under one's own control.

3. Creation of a knowledge text or a message, and its transmission has seen a sea change. Now the whole process is seamless. Original composition of an information text (in the case of structured texts), its reviewing, refereeing, editing, typesetting, printing and forwarding are done online or following assembly line production via the computer. In the case of unstructured messages, one has gained unprecedented freedom to modify messages through addition, deletion, shifting of matter, change of fonts, colour combination, insertion of graphics (possibly insertion of voice in the near future), and page design etc. directly on the screen. This freedom of action was unthinkable a couple of decades ago. The result is mushrooming growth of information and information resources - wanted and unwanted.

4. In many cases users have now the option of reading newspapers, journals and research papers online as well as in their print versions.

5. Unless one is computer literate it is becoming difficult for citizens to cope-up with the procedures connected with many of the public services and matters involving interface with governments like filing a passport application or applying for a loan or reserving a journey ticket and asking for government information. Alternative is to seek help from outside agencies, which involves expenses, time dependency and the question of reliability.

6. On the negative side are the problems associated with e-form information, such as defrauding users, conducting illegal activities, and even exporting terrorism, apart from health related concerns associated with using a computer for long times or continuous use of mobile phones.

3.4.6. Impact of Complexity on Information / Human Affairs

The term complexity is used here in its general sense to mean the state of things of a system/society caused by hefty increase in the number of its constituent elements /forces, which in turn accelerate the rate of interactions between them and results in rapid changes. It is inclusive of the sense conveyed by socio-cultural development. The effects of this change are contagious. Its impact can be felt instantaneously and globally. Three examples are given here to show the types of impact of complexity on society: government generated information, infructuous time, and medicalization of human body.

3.4.6.1. Increase in Government Information

The total stock of knowledge/information held by a society when organized mirrors the nature and composition of that society. Complexity is one of the most powerful features that changes the nature and composition of a society. Complexity increases when the elements, groups and forces making up the society increase enormously; and the interplay between them leads to generation of compound effects having impact on the whole society. In other words, a society can be considered as a system. It is a dynamic system where changes occur rapidly, their effects are widespread, and the effects of change impinge on many sectors of the society. As a result, governing a complex society becomes a challenge to the three pillars of the state, namely, legislature, executive and judiciary. The growth of complexity in the present evolving society is caused by the enormous increase in the industrial and commercial activities owing to rapid rate of scientific and technological developments and their application, impact of globalization, and the fast pace of social changes taking place. The consequences of this phenomenon are: the generation of immense quantities of governance related information and data, the increasing demand for such information from a wide range of sectors to be provided quickly, and the need to cope with the speed at which new data and information are generated and the consequent increase in the rate of obsolescence of parts of the data and information..

Growth of complexity in a society leads to increased activities over a wide range and variety of domains. A good governance of such a society calls for enactment of laws covering the new changes, besides constant reviewing of existing laws and regulations. Though the enactment of laws is an ongoing process in all societies, what is special about the new society is the scale of changes taking place and the number of people and groups that are affected. In this situation, the power of the political class and the bureaucracy gets enlarged enormously, which in turn tremendously affects citizen's domains, although suitable safeguards may be built into the system to limit the misuse of authority and information. Many of these rules and regulations carry a time element. It is not easy to draw the line separating justifiable jurisdiction of the government and the area of citizen's spheres as a number of forces and interests, both genuine and vested, are involved. Seen from a temporal angle, most of the laws and rules or amendments and repealing of acts are reactive in nature i.e., post action or preventive remedies aimed at controlling the situation caused by undesired and unforeseen effects of change. For example, the need for enacting laws to prevent cyber crimes arose only when such crimes began to occur, i.e., only after the appearance of the Internet. A consequence of increasing complexity is the increase in the functions and activities of the three pillars of the State as mentioned above, namely: a) legislature which enacts laws; b) executive, i.e. the government which enforces laws enacted by legislature, and frames required regulations and procedures; and c) judiciary which ensures that the enacted laws are judicially valid, and their enforcement by the executive branch is according to law as interpreted by courts of law. A government is an enormous machinery comprising ministries, departments, subordinate offices; government sponsored or controlled institutions and regulatory authorities, and so on. Each entity generates information both in print and electronic form. The net result is that the three branches of the state now account for massive output of information and data - an important feature of the present-day society.

A negative feature of the process of handling information in governments is that in spite of introducing checks and balances, control and authority mechanisms, arrangements for smooth flow of information, given the time lag in the transfer of information between agencies, it often happens that one branch of government may not know what the other branch is doing or may have done. This type of ignorance (genuine or made-up) raises the question of accountability. It is often seen that there are no signboards "buck stops here" in the system of governance. There is no knowing the percentage of state generated information in the total stock of information in the world and in each country. The topic is worth researching.

Particularly in the fields of governance, politics, and in certain areas of social life time is often used for its strategic value. The idea is to derive maximum benefit from the sensitive information pertaining to an issue held by a party by its strategic disclosure—time-wise or ward off an embarrassing disclosure which harms the self interest of the party or individual holding that information. It involves decisions as to what information needs to be disclosed and when; and how long a particular information item needs to be kept dormant. Behind this strategy are the principles "let the sleeping dog lie" and/or "strike when the iron is hot". Selection of time in using or disclosing a particular item of information of potential consequences can tremendously influence the outcome of a situation.

3.4.6.2. Impact on Human Time

Increasing complexity of modern life infringes on one's work time and free time by increasing the duration of infructuous time. This effect is further accentuated by increasing urbanization.

Infructuous Time : Twenty-four hours of a day available to a person can be partitioned into: a) work time, i.e. the time actually spent on the work at the work place; b) personal time or free time, i.e. the time available for spending as one desires, such as on rest, on family, on recreation ; and c) infructuous or wasted time, i.e. the time required to be spent in connection with the work or in fulfilling the free time objective, but not forming a part of the work or free time. Infructuous time cannot be avoided; can only be minimized. Examples for wasted time are: the time required to be spent on transportation to reach one's work place or time spent on waiting. In many countries now, with good information / computer networks in place many 'office tasks' are being done at home and transferred over the network. This saves commuting time, reduces to some extent traffic congestion on roads and exhaust gas pollution, and perhaps enables more time to be spent on other work or at home (a particularly attractive option for working women). This virtual employment has also enabled persons taking up multiple jobs. Thanks to ICT developments, some persons may use this sort of 'infructuous time' to communicate using mobile devices over networks or even work on laptops and handheld devices

3.4.6.3. Extension of Powerbase

Complexity leads to specialization, and specialization leads to acquisition of power (sociologists call it "structural power") by professional bodies and specialists. For example medical profession has derived enormous power through what Illich calls medicalization of society (Illich, 1975). What it means is that professional medical knowledge is imposed on population in such a way that it takes on a self-regulatory form on the part of individuals, and thereby the decision of the medical profession on how individuals should behave medically becomes a sort of scientific truth and assumes the form of disciplinary power (Foucault, 1977). This is how women have been self- influenced to accept hormone replacement treatment (HRT) for menopausal and post-menopausal use, which raises what is purely a part of natural process of ageing to the level of a disease requiring medical treatment, as if all women suffer from the ill-effects of a lowering estrogen levels.

3.4.6.4. Extension of Life Expectancy / Life Span and Actuarial Tables

Developments in health and medical fields and their application combined with economic progress in many countries have led to a significant reduction in child mortality rates, and increase in life expectancy and life span of people. One of the parameters taken into account in preparing actuarial tables (insurance) is the average life expectancy at different periods of the people in question. As life expectancy increases the actuarial tables are updated and the insurance premium is re-calculated.

A related matter is the aging of the population. Already in some countries (e.g. of Europe, North America, and Japan) the percentage of people of 70 years and above has increased significantly during the past decade or so. Provision for the health and welfare of this population is a national concern to be taken care of in the national socio-economic planning based on the collection, processing and analytical study of the related data. This is also true of India, where longevity of people has increased in the last two decades. Another current trend is that the number of senior citizens (i.e. above 60 years of age) in the total population is increasing when compared to earlier years.

3.5. INTERNET AND LANGUAGE

Advent of the Internet and developments in telecommunications have significantly affected the ways in which millions of people communicate, interact and transact business and personal affairs every day even simultaneously across continents, all this taking place at electronic speed, apparently saving considerable amount of time. These developments have also introduced very many new words and phrases and in some ways changed the way we query the Internet and talk into telecommunication devices and work with handheld communication equipment. In this section we shall briefly examine the evolution of language over time *vis a vis* the Internet and related devices.

Language Evolves Slowly Over Tme : Language change and its outcome are noticed only after some sometime. Most languages have seen revolutionary changes at different periods for different reasons. (Crystal, 2005, p.4.) observes that "the second half of the 21st century- and the 1990s in particular - a highly significant period in the history of languages due to coming together of three major trends, each global in its implications, which together have fundamentally altered the world's linguistic ecology". The three trends traced by him are: a) emergence of English as the world's first truly global language, and its consequences on English itself; b) the crisis facing a large number of languages which are currently endangered or dying, and the steps being taken to preserve and regenerate them;

and c) the radical effect of the Internet technology, which has supplemented spoken and written language with a linguistically novel medium of communication, and raised fresh questions about the way language will continue to evolve. The effects of Internet described below is largely drawn from David Crystal.

Impact of Internet : Internet language is a new stylistic variety of language marked by large-scale use of slang and jargon, and of spelling and punctuation bypassing many linguistic rules. More importantly, the Internet has become a novel medium of communication, combining the features of writing and speech. This alternative medium is generally represented by the name *computer-mediated communication* (CMC) and *electronic communication*, and its language *Netspeak*. The Internet is seen as an amalgam of television, telephone and conventional publishing which conveys the notion of a world of information in digital form and hence named *information superhighway* and *cyberspace*. Internet technology is used to carryout three broad functions, namely, the World Wide Web providing access to all types of information stored in thousands of websites throughout the world; electronic mail enabling transfer of messages between private mailboxes; and c) chat groups (social networks) which allow participants interested in a particular topic to discuss in an interactive mode in a chat room. This is entirely a new linguistic situation modulated by the psychology of working in an electronic environment tempered by features of space and temporality; the individual's attitude; and psychological aspects of communication, such as tone, body language, facial expression, and the peculiarities of language used for communication. To overcome some of the limitations of Internet mode of communication efforts are continuing in the form of formulation of codes, compilation of dictionaries to reduce ambiguity of language expressions, and inventing devices to incorporate psychological features such as tone of expression and spontaneity of response to a message. Netscape is more than a hybrid of speech and writing. It combines several properties that are unique to electronic communication and have consequences to language development. When these properties are homogenously combined with those of speech and writing, the resultant Netscape will become a genuine "new medium".

The linguistic consequences of the new medium within a language are: a) initiating change in the formal character of the languages which use it; and b) new opportunities provided by it for languages to use.

Changes in the use of formal language by Internet users ,i.e., *Internet graphology*, relate to initialisms (eg. *afaik* for 'as for as I know'), acronyms (e.g., FAQ for 'frequently asked questions'); selection and use of capital and small letters (e.g., *AltaVista*); with the tendency to prefer lower case letters; coining of new vocabulary, many of which have their roots in computer software and hardware language (e.g., *file, online, wizard, search, address, save, send, lock, crash, roaming*), and words identified with the computer user (e.g., *netizens, surfers*); and combining two or more separate words or parts of them to form a new word (e.g. *cybersex, hyperfiction, infonet, netiquette*) - all dictated by the needs or peculiarities of the worldwide Internet situation. (Crystal, 2005).

Linguistic Dispersal of Internet : Internet which started with English as its language and had dominated the networld till nineteen nineties is now overtaken by non-English languages, such as German, French, Japanese and Spanish. Asian languages like Arabic, Chinese and Hindi are emerging as major Internet languages. Factors like the need for language identity and cultural fulfillment is motivating even small and marginal languages with very limited number of speakers to create web sites and advertise their existence. A major hurdle in Internetizing non-English languages is the practical difficulties involved in developing keyboard capable of representing non-English alphabets and word ascents. Crystal points out that the UNICODE system in use today is capable of representing 94,000 characters on the screen, while the total number of written characters in all the world's languages is estimated to be about 175,000.

Indian Stage : British rule in India impacted on Indian language space in two major ways: a) Imposition of English as the sole administrative language of the country exposed educationally advanced section of Indian society to Western thoughts. It also prompted them to learn English language which served as the gateway for employment in administration; and b) kindled a new awakening among the communities speaking different native languages on the need for safeguarding their linguistic identity and fight for rightful place of the local / native language in their respective geographic domains. English language's position in India was firmly fixed when the Constitution of Independent India retained English as an official language along with other major Indian Languages. Globalization and the prominent role of the English language as the instrument of advanced technological development, and higher education coupled with urbanization have influenced all major Indian languages which is a topic for several studies. What is of importance here is to know how Indian languages are faring as Internet languages.

Several database software (commercial and non-commercial) are now Unicode compliant enabling handling of non-Roman scripts. For example, the open source free software Greenstone Digital Library Software created through collaborative effort at the University of Waikato, New Zealand, has outsourced the development of user interfaces in South-Asian languages including Indic languages, such as, Tamil, Malayalam, Telugu, Kannada, Marathi, Hindi (*see* Raghavan, 2010)

Indian language web sites are developing, though not in proportion to its readership. There are several reasons for this, which include technological (hardware and software related); dominance of English speaking urban culture in matters of development; changing life styles; slow and low spread of education, particularly computer literacy in society; and lack of financial support. Whatever growth is there, it is mostly the result of initiatives from committed individuals and enterprising groups / ventures, and bodies promoting language and literature. One result of these initiatives is the publication of online journals catering to original language texts and/ or their English translations (e,g. *Muse India, Prathilipi and Kritya*). (Raju, 2009). However, content development in native languages is lagging behind for several reasons, such as: Internet access remaining mostly an urban phenomenon; use of translated content rather than creating content in original language; relatively uneconomic nature of configuring hardware and software around one language out of the large number of Indian languages. This has hampered development of Indian language search engines (Kohli-Khandekar, 2011). Though there has been tremendous growth in audiences and revenue from language newspapers, it has not resulted in many online language newspapers. However, creation of web contents in Indian languages will now be facilitated by the Machine translation (MT) systems, a programme launched recently with funding from Technology Development for Indian Languages (TDIL). This tool based on Paninian grammar when completed will help translate Indian languages into Hindi and other Indian languages. At present three versions are available. (*Business Standard*, 2011, p.10). In the development of content in various Indian languages, and in providing them international exposure, the contribution of Wikipedia is substantial.

Wikipedia : Wikipedia has its Indian chapter and now has opened its first offshore office in India. It intends to expand its presence in India by taking a collaborative approach to create and expand local language content through local Wiki communities. It has already created over 20 Wikipedias covering all major Indian languages which add up to over 2 lakh (200000) articles, thus forming the largest collection of Indian language online content. It has not been an easy path, and had to find innovative solutions to problems posed by Buggy fonts, lack of platform independent fonts, and absence of common standard for keyboard layouts. There are yet a number of shortcomings to overcome, e.g., system incompatibility in using *Nudi* font, the standard accepted by the Karnataka Government, in preparing data input. A new factor that will have consequences for the development of vernacular content is the substantial interest shown by such companies as Google, Microsoft and Yahoo. Private companies are investing heavily on the development of tools that can help translate the web and 'localize' it. Google is encouraging Wikipedia volunteer groups to use their translation tools. It is also engaged in translating Wiki articles into Hindi. Google foresees substantial growth and benefits in the development of indic content. One of the benefits for the company is that it helps to improve their Translation Toolkit by increasing its linguistic ability. Other translation tools that work on Wikipedia include Microsoft's WikiBhasha and Yahoos's Babelfisch. However, the Indian Wikipedia community is divided on the efficacy of this 'commercial' approach to content creation (Kurup, 2011).

4

Socio-Cultural Differences

4.1. EXPRESSIONS OF TIME : CULTURAL DIFFERENCES

"Does time change? Time stays still but events change, which we interpret as time changing." (Hegde, 2011),

"Thought becomes possible only when the mind makes its passage through Time. Time is the very form of thought." (Mehta, 2011)

Cultures differ in their idea about time. Compare, for example, the European and the African concepts of time. The African culture is not one homogeneous culture, there are different cultures in Africa. Some African peoples' concepts may be more like those of the West than of some others'. The viewpoint of some African peoples and individuals today will be different, although the traditional view also pervades modern Africa.

In the European view, time moves forward, coming from behind us; time exists as an entity in itself, and it moves. "We speak of time travel in science fiction in the terms of forward in time or backward in time. In the Western view, an event is a component of time. As time moves, you must use it or lose it. If you do not use it, it is gone." On the other hand, in the African view time flows backwards, that is

" It flows toward you from the future, and the more or faster the activity, the faster time flows. Time is created, in a sense."

Time is not something in itself. Life is made up of events, defined by relationships. Time is a component of the event. "Your activity really determines the amount of time that passes. Thus the faster you work, the more time you use, because more activity is occurring, more energy is being used. If you are sitting and resting, you are conserving time. Time is not actually passing; it is simply waiting for you." (Jenkins, 2008, 2009, 2010)

In **western society and corporate cultures**, past is context. It helps shape the present and guides in making future choices. Present is determined by clock time. Past and future merge in the present. Without activity, time hangs on and stress caused by hectic activity may lead to the feeling of too little time. Future time is the present time mortgaged for future benefits for the self and family. (Whybrow, 2006) (Mito, 2007)

In **Eastern cultures** past is the time already lost; present is the time "not-lingering"; and the future is time still awaited. Mind is not in any of these times. Highest form of reality lies in "time-less time" which covers all of existence including mind.

Old Testament - Christianity, Arabic and **Hebrew**: In the Old Testament time was traditionally regarded as "a medium for the passage of predestined events" (Wikipedia).

African Cultures: In African cultures time is expressed by the terms "Sasa" and "Zamani". Sasa refers to the present period indicating a sense of immediacy, nearness, and "now-ness", and relies on social interaction, and community life represented in the natural ebb and flow of life. It is created by recognizing opportunities in nature and community and acting upon them. Sense of past is represented as events recently experienced; present as events in the process of realization; and future as events about to occur. Zamani presents a sense of binding together of all created things. Past, present and future are represented as memories of marriages, birth, war, famine, journeys and so on. One is dead only when the incident fades away from memory which may linger on for several generations.

Sami Time: One of the more interesting widely studied indigenous cultures of Europe is the Sami culture. The concept of time of these people of the **Lapland** is based on their strong relationship with Nature, the reindeer in particular. Lapland is a region in northern Fennoscandia, largely within the Arctic Circle stretching across Norway, Sweden, Finland and the Kola Peninsula (Russia).. On the North it is bounded by the Barents Sea, on the West by the Norwegian Sea and on the East by the White Sea.

Theirs is experiential knowledge "accumulated through repeated experiences of particular situations and arising from "the idea of man following the flow of Nature without altering it, and that one cannot rush Nature. The reindeer will begin to migrate when Nature dictates, not when the Sami decide that it is time. "Sami time is based on the cycles of nature, particularly the yearly cycle of the reindeer. One cannot predict the specific moment when reindeer herds migrate or ptarmigan chicks hatch; it changes from year to year, season to season. "In a culture based entirely on Nature, time is relative to physical conditions, not necessarily the position of the stars or the planet, much less the oscillations of an atom."

For these nature-dependent people, knowing when the reindeer would migrate and when the salmon would spawn is of greater importance than knowing the exact second, minute and hour of a particular moment in time. Specific days of the week were also part of the Sami calendar until the introduction of Christianity and the specific holy days associated with the religion.

The seasons are harmonized with the life-cycle - the annual behavioural pattern - of the reindeer marked into eight seasons: Spring-winter, Spring, Pre-summer, Summer, Pre-autumn, Autumn, Pre-winter, and Winter

The Sami calendar parallels more or less the western calendar, but the months indicate some event(s) in the reindeer's life as indicated in the table below.

The Weeks: The Sami concept of time also includes various names for the different weeks in a year. Only some of the week-names have survived. It is likely that the use of calendars designating weeks become a necessity after the introduction of Christianity and related church services. Some of the events are those of the Christian calendar.

Sami Calendar

Western	Sami	Meaning
January	Oððajagemánnu	New Year Month
February	Guovvamánnu	Unknown
March	Njukèamánnu	Swan Month
April	CuoKománnu	Snow Crust Month
May	Miessemánnu	Reindeer Calf Month
June	Geassemánnu	Summer Month
July	Suiodnemánnu	Hay Month
August	Borgemánnu	Molt Month
September	Èakèamánnu	Fall Month
October	Golggotmánnu	Rut Month
November	Skábmamánnu	Dark-Period Month
December	Juovlamánnu	Yule Month

The following list of Sami week and the wooden calendar, both are strong indicators of this change in the Sami culture. The names of weeks relating to Nature coincide with those of the earlier mentioned concepts of time.

Sami Weeks

Sami Name	Meaning (Week)
Bar'debei-vâk'ko	St. Bartholemew (Aug. 24th)
Birgit-vâk'ko	St. Birgit's
Dal've-vâk'ko	Winter
Gil'de-Marija-vâk'ko	Virgin Mary
Hallemes-vâk'ko	All-Saints
Mattusmes-vâk'ko	St. Matthew's
Miccamar-vâk'ko	Mid-Summer

Mikkalmes-vâk'ko	St. Michael's
Oar're-vâk'ko	Squirrel
Simmun-vâk'ko	St. Simon's
Sobmer-vâk'ko	Before Leaf-fall
Gássa-vâk'ko	Goose
Urba-vâk'ko	Trees Leafing

Western society's impact: The Sami calendar began to give equal importance to the religious cycle of the year and to Nature's cycle, indicating a move towards the assimilation of the Sami culture. As tax collectors, traders and missionaries moved into Sami land the weekly calendars became a necessity - a more fixed timetable than the natural calendar of the Sami.

The Sami "wooden calendar is also divided into weeks and it serves as an easily transportable method for telling time. Typically fabricated from wood or reindeer antler, this calendar was used to keep track of both natural phenomena and religious occurrences. Written in the runic alphabet, the wooden calendar was a useful tool in helping to preserve the balance between Nature and religion as they denoted events in both realms. These calendars date back to the mid-1800s and are another indication of the impact of Christianity upon the Sami."

"In the drawings of the Sami wooden calendar, there is much detail and attention given to the intricacies of their way of telling time. Crosses denote days of religious significance while fish and leaf sprigs denote various events in nature. By keeping track of time in this way, the Sami could easily refer back to earlier times of the year in order to try to predict when the fishing season would be most bountiful or whether or not spring would arrive early or not. This sensitivity towards the ways of the natural world was and is mandatory for the reindeer herders, fishermen and farmers alike."

South American Cultures: How the meaning attached to the concept of time varies with different cultures is briefly described below. (McWhorter, 2009):

1. "ahora". This Spanish word used in Costa Rica (English

equivalent of "now") conveys the meaning something urgent. Costa Ricans use ahora at three different levels, namely (a) ahora really meaning later; (b) ahorita to mean now but in a little bit; and (c) "ya" meaning NOW.

Although ahora, ahorita and ya correspond to the English words "later", "in a bit" and "now" respectively, the Spanish usage indicates the English meaning of generalized "now" having a time frame of long latitude. Costa Ricans have another term "aharo mismo", which also means "now", but is used in a more formal context. Costa Ricans appear to have a less definite sense of time and it is so flexible that an Englishman or an Irish have no use for it. However, ahora, ahorita, and ya can be expressed in English with the terms "soon", "now", and "right now", and with the help of context.

2. "tico time". Costa Ricans use this term jokingly to jibe at fellow Costa Ricans' unpunctual time sense.

North America - Hopi Tribe: The language of the Hopi Indian tribe mostly living in northeast Arizona due to non existence of verb tenses misses the concept of time. "The closest that the Hopi language comes to a sense of time are two words in the entire language: one meaning "sooner" and another meaning "later (Le Lionnais, 1960).

Vietnam and Indonesia: The Vietnamese "gio*da^y thun and the Indonesian "jam karet", having the literal meaning: rubber band time, express time elasticity in Vietnamese/Indonesian usage referring to unpunctuality (8.00 could mean 8:45 or even 9:00). In contrast to the South American cultures mentioned above, the Vietnamese language has only two levels of "now": (a) "ba^y gio*\ meaning now but not immediately, and (b) " ngay ba^y gio*\ meaning NOW.

Afghanistan: For certain nomadic tribes of Afghanistan and Iran the concept of time is identified with the beginning and end of seasons. Their movement in search of richer and abundant grasslands for their animals is prompted by the beginning of spring and end of summer.

Cultures of India: Mythology and Philosophy: Hajime Nakamura (1993) in his essay The Notion of Time in India (summary by Angie Weicekauskas) says that the concept of Time in Indian thought is quite different from that of the West. "In Indian thought, time, like other phenomena, is conceived statically rather than dynamically.... While it is recognized that the things of this world are always moving and changing, the substance of things is seen as basically unchanging, implying that its underlying reality (unchanging universal) is unaffected by the ceaseless flux (of time).... Indian thought places a high value on universality, and hence the connection between this, and the static conception of phenomena. 'The one remains, the many change and flee.' ... This static conception of time permeates Indian thought, in the very forms of language itself, conditioning all philosophical thinking." Nakamura provides examples from the Sanskrit language.

Nakamura adds: "The persistent Indian conception of a transcendent reality as more important than the phenomenal world it underlies and sustains results in a kind of paralysis of the individual's sensitivity to time, if we understand "time" to mean the passage and flow of specific events in our experience......Indian thought may show an intense preoccupation with other, more metaphysical senses of time; in the Vedic period time was seen as the fundamental principle of the universe, "Time, the steed, (who) runs with seven reins, thousand-eyed, ageless, rich in seed. The seers, thinking holy thoughts, mount him, all the worlds are his wheels.... With seven wheels does this Time ride, seven navels has he, immortality is his axle.... Time, the first god, now hastens onward". But this is hardly the time in which human beings carry on their common concerns." .... "Indian philosophers in general replace the concept of Becoming by three aspects of temporal existence: Appearance, Extinction, and Continuance. All three states are clearly conceived as static. They are referred to early in the Upanishads and are generally accepted by the orthodox schools of Brahmanism and Jainism."

Again, in respect of Time *vis a vis* ancient history Nakamura writes: "We find the Indian conception of history very different from our own. Indian books of history ... are tinged with a fantastic and legendary color. They are not products of historical science but rather works of art, usually written in verse. Indians are not satisfied with the simple description of facts in the language of daily use. They beautify the past and try to idealize it. They ignore precise figures, exact sequences of events, and other details of time and place." But "It would be incorrect to infer that Indian people have no concept of abstract time. On the contrary, the view of the uncertainty and transiency of life which is at the center of both Buddhism and Jainism demonstrates that they understand from their heartfelt experience the changing phases of the world."

Romila Thapar, Historian, Professor Emeritus, Jawaharlal Nehru University, New Delhi, India, in her Hudson Memorial Lecture (Tapar, 2010) on 'Cyclic time and linear time in ancient India' and in her book *Time as a metaphor of history: Early India,* (1996) emphasized that "Ancient India had both the cyclic time and linear time concepts". She argued that the view that the cyclic time prevented the evolution of a sense of history is not correct, pointing out that "Early India also had the concept of linear time, implicit in a variety of forms. The linear time was associated with a sense of history and the two concepts (cyclical and linear) intersect on some occasions. Explaining further, she said "the cyclic time might be described as cosmological. The linear time, (historical time), was functional and dependent on human activity. Distinctions could be made between the cosmological time and the historical time but the degree of separation or overlap would vary depending on how history or the past was perceived. Though both were carefully constructed, the cosmological time was a conscious fantasy of time and reflective of authors and their mythologies, whereas the historical time reflected more manageable concerns. Further, "The two concepts did not exhaust variations of time. The simultaneous use of more than one form of time indicated that different segments of society viewed their past in different ways. If time were to be seen as a metaphor of history, many dimensions in historical intersections required to be explored."

India with its five thousand years of heritage has generated a very rich corpus of time-related concepts. The contributions of its philosophies, religions, cultures and traditions to this collection are unique in the world. This can be seen from the fact that Tamil lists more than 800 concepts associated with time covering almost all aspects of life. This is equally true, to varying degrees, of other cultural groups of India - Hindustani, Punjabi, Gujarathi, Marathi, Kannadiga, Malayali, Andhra, Bengali, Assami, and a number of indigenous tribal communities. Reasons for this rich growth and accumulation of temporal concepts are many, and they include:

1. The nature, location and vastness of the country having diversified geography, climate, sea, rivers and mountains, fauna and flora which serve as a fertile platform capable of nourishing highly philosophic as well as mundane ways of thinking, and styles of living;
2. Capacity to assimilate features of different political systems accounted by Aryan and pre-Aryan (Harappa – Mohanjadaro and Dravidian cultures), Greek adventure followed by Persian and Mongolian conquests, and ending with the Portuguese, the French and the British colonial rule;
3. Indian culture and society has been a host to different ideas and customs flowing form all major religions of the world - Hinduism, Buddhism, Jainism, Zoroastrianism, Islam, Christianity, and Judaism;
4. Development and existence of numerous languages and dialects, rich in cultural terms, enriching the polity;
5. Adoption of the English language exposing Indian society to concepts developed in the Western World; and
6. Capacity of the Indian mind to synthesize, adopt and adapt different concepts into life course.

It is in this milieu embedded in time Indian culture has advanced and life course has been shaped. Modern Hindu/Indian culture is inspired by its ancient philosophy and Vedic culture and the thinking and ways of living of millions of Indians are governed by them. Smrutis prescribe the social norms and cultural practices to be followed. The essence of Bhagawad Gita, the sacred text of the Hindus, can smoothen a frayed Hindu mind; even applied to interpret modern business/management practices. Buddhism and Jainism were born as alternative religions with reforming platforms. All the three have contributed most to the development of knowledge in India and abroad - philosophical, religious, social, cultural, political and economic. Hence, the study of Indian time concepts has to take into consideration its frame of philosophy and religion and the languages of expressions (Pali, Sanskrit, Tamil and other Indian languages) through which they are communicated.

Indian Mythology : Celebrations, such as the New Year marks the passage of time; in the case of the Gregorian calendar it marks the passage of 12 months and the commencement of another. In the cosmic time frame human existence is just a microscopic proportion. The Indus Valley civilization is gauged to be about 5000 years old. The Tamil Sangam literature has documented, among other things, 2000 years old places of worship (temples). About 1400 years ago the Bhakti movement in South India documented the existence of temples. The Ellora (cave) temple is of an even earlier date; so also the Tanjavoor Periya Kovil (temple), the Hoysala monuments and other temples that are also architectural monuments are over a 1000 years old. And over four centuries ago the emperors of Vijayanagar created great monuments all over south India.

The smallest unit of time is a *kaashta* which is 18 times the length of time to blink an eyelid. Ten *kaashtas* make a *kshanam*, 12 *kshanams* make up a *muhoortam*, and 60 *muhoortams* make up a day; 30 days constitute a month and 3 months make up a *ritu*, and 12 months make up a human year.

In the realm of departed souls - the *pitrus* - a human month equals the length of a day. The brighter half of a lunar month is the *pitru's* day time and the darker half constitute their night. In the realm of the Gods or *Devas*, a human year is equivalent to a single day. The brighter half of the year or *uttarayanam* makes up the day time hours of the *Devas* and *dakshinayanam* or the darker half constitute the night time hours.

A *yuga* (epoch) of 1200 *Deva* years (1200x360=432000 human years) constitute the *kaliyuga* or the present epoch; the preceding *dwapara yuga* was twice the length of the *kali yuga* that is, 2400 Deva years; the 3600 Deva years preceding the *dwapara yuga* made up the *Treta yuga*; and 4800 Devas years preceding the *dwapara yuga* made up the *Krita yuga*

A cycle of 4 yugas is called *catur yugas*. A cycle of *catur yugas* lasts for 12,000 Deva years or 12,000 x 360 i.e. 4,320,000 human years.

In Indian mythology, these time measurements may be correlated with the process of creation thus:

Brahma is referred to as the creator. A thousand *catur yugas* make up the daylight hours of a single day of Bhrahma's life. Another thousand make up the night time of a single day of *Brahma*. Thus, a single day in Bhrahma's life spans 2000 x 4,320,000 = 8,640,000,000 human years; 360 such days, each lasting 8.6 billion years constitute a year in *Brahma's* life, which lasts for a 100 Brahma years. At the end of one Bhrahma's life, another cycle starts. A Brahma's life is also known as a Para. Each half param is a *parardham*. It is said that we are currently living in the 2nd half of the life of the present Bhrahma!

Vedic rituals : The time-frame in which ritual is being performed is specified both in macro and in micro terms, the term '*dviteeya paraardhe*' (the second half of Bhrahma's term) is stated. The reference point here is the moment when Brahma began the creation of the Universe. Krishnaswami writes, "When we say '*dviteeya paraardhe*', which Bhrahma are we referring to? How many Bhrahmas have preceded the current one? This specification is non existent in vedic mantras. Since the whole process is cyclical, with one Bhrahma commencing when another completes, and with this process repeating forever, there may not be any significance in stating the position of Bhrahma. In other words, in a cyclical concept of time every starting point will have to be also an ending point. If time is postulated as being linear and unidirectional there will have to be an absolute starting point for time. This cyclical nature of time in Indian mythology refers to time as that without a beginning or '*anaadI*'

Another measure of time is *Kalpa.* A kalpam or an epoch is made up of 14 *manvantaras* and each *manvantara* spans 71 *caturyugas.* The fourteen *manvantaras* are respectively *swayambhuva, sawosisha, audhama, thaamasa, raivatha, sakshusha, vaivasvata, savarni, daaksha savarni, bhramha savarni, dharma savarni, rudra savarni, rouchya and bowdhya.* The present *kaliyuga* is the 28th in the present *Vaivaswata manvantara.* Each *Brahma's* term lasts for 7 *kalpas.* The current period in time is said to belong to the *sweta varaaha kalpam,* which is in the second half of the life of Brahma.

The puranas are named after *kalpas,* thus *matsya kalpa, koorma kalpa, lakshmi kalpa, sweta varaaha kalpa, shiva kalpa, bhrahma kalpa, vishnu kalpa* etc.

Sankalpam : *Vedic* mantras pin point the time of performance of a ritual - by narrowing down from *dwiteeya paraardhe* (in the 2nd half of the term of *Brahma*), *Sweta varaaha kalpe* (in the *kalpa sweta varaaha*), *Vaivasvata manvantare* (in the 7th *manvantaram*), *Kaliyuge* (in the Kali epoch) - through the finer details such as the name of the current year, month etc.

Other frames for kAlam (time, duration)

kaNam = Shortest duration of time as measured by a snap with the fingers.

ilaku = One of 10 varieties of kAlam (time, duration) which consists of 16384 *kaNam*

kaNam. kAkapatam = One of ten varieties of kAlam which consists of 65536 *kaNam*, etc.

. patam = nAzhikai = Indian hour of 24 minutes

4.2. CONSEQUENCES OF CULTURAL DIFFERENCES

As the concept of time is affected by philosophical and cultural differences, in interactive communication among people of different cultural groups misunderstanding and conflicts arise. This is more so in the present day global environment where interaction between peoples of different countries and between people with different socio-cultural time concepts is on the increase. For example, the term 'Sabbath' is Saturday of every week that is holy and meant for rest in Jewish religion and Israeli nation, whereas in other cultures it may mean taking a day or days off, not necessarily the 6th day of the week. In academic context Sabbath means taking off from work up to a year for pursuing, say, academic/research work, travel, or leisure. One can imagine the misunderstandings that different socio-cultural time concepts can create in the minds of people belonging to different cultural backgrounds participating in an interaction, unless the meanings attached to the different 'time term' are understood. Even within a vast cultural entity like Hindus misunderstanding arises due to differences in meaning of time caused by cultural differences. Consider, for example, the concept of time in 'New year'. While for most of the sub-cultural entities like Telugu/Andhras (*Ugadi*), Kannadigas (*Ugadi*), Maharashtrians (*Gudi Padva*), and Tamilians (*Puttaandu*) the new year begins on a day in the 1st week of *Chittra* (14 April); for Gujarathis (*Bestu*) and the business community it begins on the day before *Deepavali* usually in the month of October. (Krishnaswamy, http://www.templenet.com/beliefs/concept_of_time.htm)

Question : Why do some people seem to live their lives by the clock while others have a more laid-back approach to time?

This may have to do with the values and expectations of the particular cultural group they relate to, grow up in and are conditioned by. In this regard there are differences between various "ethnic" groups, social groups, regional groups and even clans and families. So differences will be discernible even between different family groups or lineages in what might otherwise be considered as one ethnic or social group.

Some families or ethnic groups value abstract structures and schedules more than others. Some societies are highly structured. How things are done and how long it is expected for something to get done depends on the overall context of the event, the group expectations and *how much value* is put on time efficiency. All cultures have some concept of efficiency, but getting things done quickly is a value only in some mechanically-oriented or technological societies, and task-oriented or production-oriented situations.

Urban settings in every culture tend to be more "time-conscious" in regard to artificial and mechanical "clock time" than rural societies ("traditional" societies). The latter tend to be more relationship and community oriented, and values related to interpersonal relationships or group identity are given greater value than tasks or production, or related technological or abstract demands.

In task-oriented linear-thinking societies time is a commodity, whereas in traditional and more rural societies, which are generally oral-relational in character, time is a function of one's relationships or the cycles of life. (Jenkins, 2003). Viewing through another frame, Zahorian presents another picture of time in different cultures (Zahorian, 2008).

4.3. PROPHESIES ABOUT THE WORLD: PEEP INTO FUTURE

Prophesying is about predicting the end of an existing world, which has reached the height of debasement and disorder to be replaced by an original order by restoring the old values and mode of living. It is a time-related religious-social concept. Prophesying is shaped by the socio-cultural traditions binding each community/ society. Whether there is any scientific basis for the validity of prophesies or not the fact remains that a large number of people all over the world either believe or are curious about prophesies, and that is why prophesies continue to be made and flourish at least as news items.

In Hinduism, "*dharma samstapanarthaya sambhavami yuge yuge*", which in the English version means that "(god) will appear every *yuga* (epoch) to restore dharma in this world". This is done by first destroying the existing order when it reaches the height of *adharma* (disorder), and then reestablishing the original order. The destruction is caused by "*pralaya*", i.e. the great deluge. Hinduism recognizes four *yugas*. Some of the sects within Hindusim have their own prophesies. For example, *Vachana Sahithya* of the Veerasaiva sect contains the prophesy (*kaalajnana*) about the reappearance of Basavanna.

Other Eastern Religions/Cultures: For Christian prophets it is Armageddon. There are similar prophesies for Buddhists and Jews.

North American Indian Tribes. The prophesies of the Native American Indian tribes are rooted in their concept of time and traditions, which can be seen from the following brief description.

North American Indian tribes experience time as cyclic. This cycle is made-up of four stages such as the seasons: spring, summer, autumn, and winter, which together constitute a hoop. At the centre of the hoop is a still timeliness, the eternal present around which the cycles revolve, and form the genesis of prophesies. In contrast, the time experienced in the Western culture is linear made-up of past corresponding to a beginning, present representing current, and a future corresponding to an end. Here, time is history. These contrasting experiencing of time represent two different worldviews. A notable feature of these world views of time is how prophesies are featured, for prophesizing is a common indulgence in all cultures - ancient and contemporary.

Salient features of Native American Indian traditions briefly are:

1. Prophesies in the Native American tribes are rooted in magic, myth, metaphor, imagination, and dreams and visions, whereas Western traditions in science, history and fact, and on established religious doctrines.
2. Living in harmony with nature and overall wellbeing of the community as a whole is central to the shamans (prophets) of indigenous tribes of America. Like a " dormant tree enduring winter, the (American) Indian's respect for the earth, their desire to live in harmony with nature, lives still, waiting to bloom again in the next world cycle. Only by honouring Mother earth, Indian shamans tell us, can we avert disaster."

The prophesies of Drinks Water, Black Elk, Wovoka Hopi, Rolling Thunder, Shoshone and other North American indigenous tribes people – are described in a book (Chester and Neelameghan, to be published).

4.4. COMMUNITY KNOWLEDGE, VOCABULARY, FOLKSONOMY

Social network is an Internet phenomenon, essentially networking of people via the Web. In most social networking a wide spectrum of people participate in creating, sharing and using web-based information and data content. In this act of sharing information in the public space millions of participating folks label (tag) the contents of the material they created or encountered with keywords. The collection of these keywords/ value associations is called "folksonomy". Tagging has become widespread and finds applications in such areas as social book marking and social photo sharing (Gruber, 2007). In this way, folksonomy is the common man's method of conceptualizing data, acting as an alternative to Semantic Web's "ontology". To extend the use and usefulness of a database / information resource whether on the Web or not it is helpful to find ways and means of accommodating folksonomies into information systems. There will be a large variety of folksonomies while dealing with the concepts and vocabularies of indigenous, marginalized, native communities. Much of this may be communicated orally, which need to be recorded, interpreted and even translated for wider audiences.

In community knowledge, formation of concepts and relationships among concepts are overwhelmingly influenced by the socio-cultural, spiritual characteristics and the needs of daily life of the people of the community (Dalrymple, 2009). Prema Manmadhan (2011) writes: 'The Karinthalakoottam songs are an oral tradition, the language archaic and dialect primitive. But it is not just a folk song group; they document rural knowledge passed on from generation to generation. The life patterns and naming practices of early Tamils – the "kurinji lore" described in a later part illustrates how concepts originate, and one basic concept influences the generation of other concepts.

It is interesting as Atran (2008) comments "... An experiment illustrates that the same taxonomic rank is preferred for making biological inferences in two diverse populations: Lowland Maya and Midwest Americans ..." This supports a modular view of folk biology as a core domain of human knowledge and as a special player, or 'core meme,' in the selection process by which cultures evolve. Structural aspects of folk taxonomy provide people in different cultures with the built-in constraints and flexibility that allow them to understand and respond appropriately to different cultural and ecological settings. Another set of reasoning experiments shows that Maya, American folk and scientists use similarly structured taxonomies in somewhat different ways to extend their understanding of the world in the face of uncertainty".

4.4.1. Culture-Dependent Time Vocabulary

Time-related vocabulary in different cultures signifies more than mere words. They also reflect traits, like punctuality, associated with particular cultures, e.g. German punctuality. Temporal words in a language may have profound effect on the worldviews of the people speaking it (McWhorter, 2009). Since Hopi Indians do not include conjugations to mark future tense verbs, they must have different conceptions of time from that of Western culture (Whorf, 1941). However, McWhorter points out that Hopi people can still talk about the future even without a future tense, like by inserting a word meaning tomorrow into the sentence. A general feature of classical and biblical Greek is that a command meaning present instant is, by and large, expressed by using the past word (e.g. use of the command "please left the luggage here" and not "please leave the luggage here".

Hence to help overcome confusion caused by socio-cultural differences in expressing time is the need for ontology of SCTE. Wennerberg and Schulz (2010, p.64) state that "the purpose of the ontology is to explicitly and formally describe these context-dependent temporal expressions. This is done by defining appropriate hierarchical categories to which the expressions will be assigned. The ultimate goal is to be able to develop context aware temporal applications for the Web such as calendars or appointment schedulers" the yearly cycle of the reindeer. One cannot predict the specific moment when reindeer herds migrate or ptarmigan chicks hatch; it changes from year to year, season to season. As already mentioned "In a culture based entirely on Nature, time is relative to physical conditions, not necessarily the position of the stars or the planet, much less the oscillations of an atom".

5
Time in Life Course

5.1. LIFE COURSE AND TIME

Life course forms the core in a temporal perspective of social and cultural knowledge. This section is largely based on Mills exhaustive review of life course research (Mills, 2000). Taking an ontological approach, Mills reviews the temporal shaping of life course from a philosophical standpoint and research from the epistemological standpoint. Data, methods and methodology are the epistemological instruments linking empirical evidence to the social processes that make up life course.

Temporality is an inherent element in life course. Life course is shaped by how social processes, such as, the family, education, and employment, health and migration domains are structured over the individual's life span (Elder, 1974; Hogan, 1981 cited by Mills). Life events such as marriage, birth of a child, education milestones, career entry and exit, retirement, are important temporal events in one's life course. These events can change the direction of life course.

Time dimension of social and cultural life is primarily made up of natural and cosmic phenomena, and cultural, social and institutional forms which together shape human life and society. Nature imposed effects determine physiological development of human body, such as time related aging process, menstrual cycle in women, and cyclic operation of seasons dictate agricultural practices, and the like. More and more institutionally framed calendar time is determining our social, cultural and vocation practices impact on life course. Technological and medical advances, innovations and devices, such as air conditioning, contraceptives, and surrogate motherhood are reshaping cultural, social and institutional constructions in some countries. This has resulted in increased colonization of time in life course by neutralizing the effects of natural and cosmic phenomena.

5.2. DIFFERENT FORMS OF TEMPORALITY

Narayana (2010, p, 272ff) summarizes the study of the effects of temporality on social behaviour that has interested many social theorists. Scholars have adopted different approaches to understand the effects of time on individuals and groups. Impact of time has been defined by theorists in the following ways: "social synchronizing and integrative function of time (Durkheim); commoditization, economic dimension and role of the state in time (Karl Marx); rationalization of time (Weber); the past is continuously re-created and reformulated as the present unfolds (Mead, 1932; 1980); a multitude of times , and not a single time, interpenetrate and permeate our daily lives" (Adam, 1990; 1995); persistently interacting multiple life worlds, where the past and future simultaneously permeate the present rhythms of the human body, planets, seasons and routines centered upon clocks and calendars (Berger and Luckmann; Schutz and Luckmann, 1973); synchronic (interdependent) and diachronic (dynamic) systems differentiate themselves from the environment and that the past and future are mere horizons of the simultaneously occurring present (Luhmann, 1977;1982); the future is merely the 'extended present' (Nowotny, 1994); and by planning our life course we seek to 'colonize the future' (Giddens, 1981; 1984; 1991; 1994); the concept of lifelines and recurrent time-space paths that give rise to world lines (H=gerstrand) (Mills, 2000). (Gerth,and Mills, 1958).

5.2.1. Natural and Cosmic time

Time operates through relatively fixed, repetitive or cyclical rhythm of the natural environment, such as the lunar and seasonal cycles. Natural time is accounted for by the influence of natural factors like temperature, latitude, photoperiod, and precipitation (monsoon cycles). How natural time affects human life can be seen from the following examples: the *temperarure* (in Canada) which is regulated by the natural cycle, and the *latitude* (of Scotland) which determines the *luminosity* experienced by individuals influences fertility, and thus affects sexual behaviour and conception in the two countries. However, as already mentioned, colonization of time made possible by technological and medical devices, has reduced the importance of natural time in decision areas like begetting a child and its birth period, agricultural practices (e.g. greenhouse crop), and work environment.

5.2.2. Biological and Psychological Effects of Time on Human Development

Temporality is an internal clock of the individual, which has an influence over the biological and psychological development of an individual. Biological development relates to physiological growth and maturation through life stages. The biological time follows a predictable pattern as it moves through its own physical timetable. The clock begins at birth and proceeds along maturation, onset of menopause in women followed by progressive cell degeneration resulting in aging, decline of health and eventual death. Nature has created differences in the development of male and female bodies. Reproductive clock of women has been the subject of many life history studies of fertility.

Circadian Rhythm is a biological process observed in plants, animals, fungi and cyanobacteria. It manifests "as an endogenous, entrainable oscillation rhythm of about24 hours. These rhythms are driven by (or composed of) a circadian clock. The term 'circadian' is derived from the Latin 'circa' meaning "around" or "approximately", and 'diem' of 'dies' meaning "day". Chronobiology is the formal study of biological temporal rhythms, such as daily, tidal, weekly, seasonal, and annual rhythms. Although circadian rhythms are built-in and self-sustained (endogenous), they are adjusted (entrained) to the local environment by external clues called zeitgebers, the most important of which is daylight (http://en.wikipedia.org/wiki/Circadian rhythm).

Biological time is increasingly colonized by socially constructed and technologically based perceptions of time. The relationship between biological and humanly devised time-tables is also the topic of study in Chronobiology. Colonization of human body by technological means, such as the use of contraceptives, by interrupting the natural temporal cycles of menstruation has empowered women around the world to control conception and sexual behaviour. There is also a socio-cognitive aspect attached to the biological phenomenon. An example is the social norms imposed on women during the menstrual period of three days in Brahmin households in India, in earlier times. Menstrual time was considered 'sacred', and the woman's body during that period as evil and unhygienic, and therefore confining such women to live in seclusion during that period. With further advances in science and technology social construction of biological time spreads. This is likely to disrupt the natural course of events to such an extent that more and more areas of everyday life will come under the influence of socially constructed biological time, which will have enormous consequences on the construction of life course.

Irrespective of the effects of biological time on women, it is the social and cultural ethos of the particular culture that influence decisions relating to age of marriage, selection of spouse, begetting children, the nature of husband-wife relationship, man-woman cohabitation, career, life-styles, and such other items in life. Here are a few examples: the institution of marriage is getting replaced by cohabitation by consent in many North European societies. More and more woman exercise their right to make individual choice regarding selection of a mate and marriage – the choice made on the basis of her perception of psychological readiness, choosing a career and so on. In India arranged marriages are still the norms. Marriage of girls at an early age are still common mostly in the northern states of Rajasthan and Uttar Pradesh in spite of the statute fixing the minimum marriage age for girls as 18 years. Even after the marriage the bride remains in her parental home till the completion of 'gauna' ceremony, which signifies the onset of menstruation and entrance of womanhood (Mukherjee, Bhattacharya and Singh, 1996). Thus, the acting of biological time is artificially interlinked with the social, cultural and historical practices of a society

5.2.3. Historical Time

Historical time is defined by the particular socio-cultural, economic, physical, and political configuration of an individual's context (De Bruijn, 1999). This along with influences of individual and family or household-based time clocks affect programming of life course. Historical events, such as war, invasion or sanctions, famine or surplus, cosmic and ecological catastrophes, ... plague or advanced health, shifts in ideology such as fundamentalism, capitalism or communism, and economic depression or boom are seen to shape a cohort, period or generation. Historical time is also used in geological stratification of the earth into different periods in the evolution of the planet and to study development of civilizations and different cultures.

5.2.4. Cultural and Social Time

The cultural and social construction of reality shapes institutional calendars which are to regulate everyday life. Behaviour in everyday life in turn makes up the social practices that construct the life course and social-cultural structure itself (Giddens, 1981, 1984). Although the geography of regions, biological life cycle and the cosmic cycle of days and seasons are generally constant, differentiation in the construction of cultural and social time stems from how different cultures transform this social space and time to reflect unique ontological conceptions.

Social Time : It refers to norms, values, rules and responsibilities about the expected time that a life event ought to occur (Adam, 1995; Zerubavel, 1981). Subsumed under social time are incidences, duration, and sequence of roles, and relevant expectations and beliefs based on age, which regulates the life course (Elder 1994). Institutional calendars constructed on the principles of cosmic events and natural cycles prescribe the appropriate time/period for conducting religious, cultural, social and vocational activities constituting life course.

Trompenaars and Hampden-Turner (1996) proposed a model of culture with seven dimensions. There are five orientations of the ways in which people deal with each other. The seven dimensions are:

"1. Universalism vs. particularism (*What is more important, rules or relationships?*)
2. Individualism vs. collectivism (communitarianism) (*Do we function in a group or as individuals?*)
3. Neutral vs. emotional (*Do we display our emotions?*)
4. Specific vs. diffuse (*How separate we keep our private and working lives*)
5. Achievement vs. ascription (*Do we have to prove ourselves to receive status or is it given to us?*)
6. Sequential vs. synchronic (*Do we do things one at a time or several things at once?*)
7. Internal vs. external control (*Do we control our environment or are we controlled by it?*)"

Each orientation and choice may have differing time involvement and management issue

5.2.5. Institutional Time / Calendars

Institutional time, viewed through institutional calendars, is defined as the time-tables produced via the adaptation and life span of institutions. Institutions are embedded within and produced by historical, social and cultural contexts. Institutions include the family, religious, education, work/production, political bodies, government, and gender divisions. These institutions are often accompanied by seemingly innocent calendars that in fact often regulate everyday life and thus most behaviour in the related life course. Furthermore, due to an individual's position in these institutions, such as being male or female and worker or non-worker, she or he experiences time in a different way. In addition to general calendars like Hindu (solar and lunar), Mayan, Chinese, Julian, and Arabic (lunar) calendars, timetables are also made for individual spheres, such as family, education, work, and sports. Globalization and inter-country exchanges have necessitated the local sports calendars to be aligned and adjusted to the corresponding timings of national and international sports events.

Historically inventors of calendars had in mind the purpose of serving the religious needs, such as determining appropriate days and time for performing rituals among Hindus, beginning and ending of fast (kutub-E-Ramzan) among Muslims, and Sabbath among Christians. Later, calendars included social, cultural and secular features such as auspicious and inauspicious time for betrothal, conducting a marriage, baptizing, starting a business or other ventures and also incorporated dates of festivals, beginning of New Year day, appearance of rainfall, and so on, which are intimately connected with life course and management. It is amazing how Hindu 'panchanga (calendar) has divided the pattern of annual rainfall into 27 parts (each part named after 27 stars) and indicating their precise onset. Equally amazing is the awareness of this knowledge by farming communities and using it to begin agricultural practices, such as sowing and harvesting.

Harvest Era : *pacali* is the revenue or harvest era, beginning from 591 A.D., and instituted by the Mughal king Akbar in 1555 A.D

Institution of the Family and Family Time : A social institution of great significance to the evolution of life course is the invention of the institution of family.

Family Time is the time shared by members of a family in a household. Family time may be further split into spousal-time, parent-child time, and time for family (kith and kin) gathering time. The amount of time and the nature of its sharing depend on age of members, comradeship or hostility between members. Joint-family institution prevalent in India which has been declining in recent times is a unique system of shared time. Traditional system of family is breaking down in many societies, particularly in the West and in urban India due to rapid industrialization and changing perception of time values. It is being replaced by more and more self-centered, right-conscious, and matter of fact "nuclear" families. Family time, properly utilized, serves as a great social cementer of values and relationships bringing greater harmony and wisdom to the family.

Educational Calendars : These constitute the first stages in the socialization of children to adapt to the norms of society to which they belong. Educational time is also to prepare the young for entering into adulthood and career life.

Work or production calendars : This is to help individuals schedule their daily activities, chart out time for family, vocation and such other engagements. Given the present day hectic phase of life, time-management has become a critical element in an individual's life course.

Time in movies : In a recent write up "Moving with Time' by Parvati Nayar (2011) says "Movies made on the concept of Time have been selling points simply because alternative ways of looking at Time ticks" She gives a number of examples of movies (mostly western)

5.3. PATTERN OF TIME ALLOCATION

Hall (2011) divides time allocation by individuals in their interactions with people into monochronic and polychronic time depending on the culture. In monochromic cultures, prevailing mostly in the acute time conscious USA, Japan and Germany, the individual attends to one-thing-at-time and is linear. It is characterized by a high degree of scheduling, punctuality, and is built around promptness in meeting obligations and appointments. Polychromic time is characterized by an individual attending to several things and people without insisting on a queue or precise schedule, which means attaching greater value to human relationships than strict business type scheduling time. Polychronic-time is a feature of Mediterranean and Colonial-Iberian-Indian cultures. On an experiential level, these two cultural time orientations may be represented by what we call "Through Time"- monochronic – and "In Time" – polychronic – time lines. (Hall cited by Palhegyi, 2006).

5.3.1. Gender Oriented Calendars

Female and male calendars, or Women's time refers to the disparate temporal experience and use of time by women and men. The most obvious difference in time use by men and women is reproductive or 'procreation' time (Tobler, 1996). Other differences relate to the perception and use of time over a wide range of issues that includes employment, free time, private time, and responsibilities. Employment issues include temperament considered more suited to women, such as nursing, air-hostess, and receptionist tasks (which are disputed by Woman-rights groups); discriminatory wage structures, and hours of labour. Private time is "deliberately designed to prevent, or at least discourage, the formation of human contact and to separate people from one another" (Zerubavel, 1981). There is also the sensitive issue of identifying women's time with house-making. With the pressure brought by woman-rights activists and vote-catching needs of politicians currently biased structure of women's time is gradually getting normalized, and the gender oriented calendar is being reconstituted. This also applies to educated and employed strata of urban India.

5.3.2. Time in Business Cultures

In the earlier leisurely agrarian period of modern history concept of time and thereby all activities were bound by cyclic seasons of the year and time of day. The notions about time in the industrial and post-industrial era have undergone a sea change particularly in Western societies. Within these societies importance attached to time varies significantly. In developing societies importance attached to time, barring a few areas having international trade consequences, time is very flexible. Kluckholm and Strodtbeck (cited by Siete) (Gutterman, 2011) group cultures time-wise into:

1. Present-oriented, with little attention given to what has happened in the past and what the future will bring. Past is seen as unimportant while future is seen as vague and uncertain;
2. Past-oriented cultures which are swayed by traditions linked

to their ancestors, family, traditionalism and aristocracy. The present is tried to be maintained; and

3. Future-oriented cultures bank on changes that enhance a more desirable development in economic and social scales.

In business and industry managing time is to help beat competition. Setting schedule of time for each task and abiding by it is a prime concern in industry and trade. Well-managed firms pre-plan every function in the enterprise from production and marketing to accounting and staffing with detailed sub-plans for each function. They also make time-wise plans to achieve short-term goals and long-term goals. To tackle any emergency situation that may arise while implementing the plan and therefore, contingency plans are made in advance. Again globalization and expanding international business activities supported by networking and networks of networks has enabled corporations to set up branches and subsidiaries in many countries (multinationals); and also or alternatively get contract work done in other (usually developing / emerging) countries at a much lower cost (business process outsourcing). An important feature of such transactions is the time differential (ranging from 5-10 hours or more between the industrialized countries of the West and developing countries in the East; it would appear that work related to a project runs almost continuously. This international sharing of work enables shortening of the project/product time cycle.

Time value of past, present and future business is influenced by respective national culture. Finns and Germans emphasize on present which is clearly linked with the past and future in similar proportions, where as France emphasizes on present and their long history. In China present, past and future are seen as independent and not overlapping periods carrying equal importance. 'Time horizon' presented by Cottle (http://www.via-web.de/time-concept/); also Time perception in France, Germany and Poland (Koslu et al, 2003) is a helpful concept which describes the nature of planning and business strategies in different cultures. Viewed through time horizon standard, it is apparent that France, China and Finland go for long-term plans, whereas Germany and America opt for short-term plans decided by the nature of the market, time-wise expectations regarding returns on investment.

5.4. TIME PACE IN DIFFERENT CULTURES

Pace of time followed is a characteristic of one's life-style, which is also true of the culture to which he or she belongs. The nature of the time pace followed by a society is molded by the value attached to it. It varies from culture to culture. Broadly the value attached to time pace in various cultures may be divided into: a) active or hectic time dictating the way of life; b) passive or slow time governing the way of life, and c) lacking in the concept of time

Hectic Time : The Japanese and the Americans live the fastest pace. Success, achievement and wealth creation is sine-qua-non with squeezing every bit of return from time. It controls the everyday lives of most American people and organizations run by deadlines. For many Americans "the 'free moments' that once glued a busy life together have almost disappeared" (Whybrow, 2006, p.15).

Slow time : It means the slow or leisurely phase by which individuals and organizations work. A slow moving society does not attach the same importance to keeping punctuality of time in attending to work or honouring engagements, and efficiency in accomplishing tasks, which is in contrast to American and some of the European cultures like the German. Life is made to move at a leisurely pace. People seem to make an exception to the slow time norms when time is related to religious and ritual connected events, at least in India. Does time run relatively slowly in most of the developing countries? However this is changing fast at least in certain segments of activities (e.g. software world) under the impact of rapid industrialization, greater urbanization, and globalization.

Cultures without Time Concept : For the *Piraha* tribe living in the Amazon rainforest with their limited language of humming and whistling the concept of time does not exist (Davies, 2000). Nonexistence of specific religious beliefs, ancestor or hero worship, and concept of numbers have made the idea of time irrelevant. They seem to live every moment of their lives without the concern of past and future.

Every society abounds with popular sayings about time that are full of insights having greater lasting value to its members. Such an unique message culled out of observations, life-course experiences and inference when expressed in a simple and compressed form make a proverb or a saying or a quotation. A few of the time-related proverbs are given in a later section..

Time Preference or Safeguarding One's Future Interests : Most human beings are worried about their future state of affairs at some stage or other in their lives. This is due to several reasons, the prominent among them being the uncertainties surrounding human life and living. What change time may cause in one's future, no one is sure. But they are aware that they pertain to the issues relating to health, economic means to support themselves in coming times, and future wellbeing of their dear ones, such as family members. In some cultures this concern is also linked with the life after death or rebirth (*punar janma* among Hindus). A general belief is that by one's predetermined current initiatives he or she will be in a position to safeguard the desired future life-course. This is equally true of institutions and organizations.

The initiatives taken to ensure desired future include planning in advance, and insuring. Forecasting forms an essential element in such initiatives. It may be based on commonsense perception of the likely future situation in the case of ordinary people or sophisticated statistical techniques and methods in the case of industry. Time preference, i.e. insuring in advance, is an economic choice, where one prefers to forego the present wants to gain future benefits that will follow the present sacrifice. Implied in this decision is the assumption that the future benefits will be larger than what is sacrificed. It is a tradeoff between foregoing immediate benefit or enjoyment and anticipated future benefit (Weirich, 2008). The essence of time preference is clearly conveyed by the antithesis of the proverb "a bird in the hand is worth two in the bush". In economics, time preference amounts to discounting the value of future goods. Future trading in commodities, taking a life insurance policy are examples for time preference. Opting for time preference contains the element of risk due to uncertainty (Narayana, 2010). Philosophers and economists are divided on the issue of rationality of time preference.

Planning : In all premeditated actions of men and women, planning features invariably in their thinking process that precedes action. It is linked to the immediate or long term duration. Given the complexities of modern-day world the need for planning and the areas of planning are ever enlarging. A timeframe is an essential part of any plan. A plan is a knowledge document incorporating a time and usually the location framework within which the plan is to operate. Not unlike reasoning, planning too seems to be an independently recognizable part of thinking. The timeframe of a plan may range from years, as in the case of economic plans to hours, minutes and seconds in the case of precision works and critical operations (Narayana, 2010)

Valentine (2008; 2010) writes about the planning of school timings in Denmark and Nepal so as to ensure that all the students are expected to follow one uniform calendar and allocated class timings. However, this uniformized timing has not been convenient to certain categories of children (e.g. emigrants) who have to assist elders or parents in certain tasks during certain hours, especially in the mornings. These children in a sense got 'excluded' from the mainstream.

Life insurance business, including preparation of actuarial tables is mainly based on the average life expectancy of people of different age-groups, health status, people's occupations, etc., in different countries

5.5. PROVERBS AS REFLECTORS OF WISDOM IN TIME-COURSE

A proverb, according to general dictionary meaning, is a short familiar sentence expressing a supposed truth or moral lesson or timeless truth. Proverbs are culled out of life experiences of people or from observations of incidents and situations occurring in a given society/cultural group. It is also considered as the epithet of wisdom.

The truth conveyed by proverbs may have universal appeal or limited to a particular society or time. Proverbs are a part of the heritage of most cultures/languages. Some proverbs have universal appeal, and hence appear in more than one language, using different expressions depending on the heritage and style of the language. However, sayings and quotations stand on a slightly different footing. These are statements picked up from authoritative texts, such as the Vedas, the Bible or unique / insightful pronouncements made by renowned persons bestowed with deep knowledge, experience and achievement in the particular domain. These do not seem to have the same popular or cultural backing as the proverbs, but have uniqueness. Several proverbs convey messages relating to time. As there are compendia of proverbs, here we give a few examples in English and some Indian languages.

5.5.1. Time-Related English Language Proverbs

1. "A stitch in time saves nine." This English language proverb shows the importance of taking necessary action at the appropriate time in advance to prevent the occurrence of later trouble in the system or relations. The proverb applies to numerous activities of daily life and human interactions. The truth of this proverb is at the heart of all preventive actions, such as preventive maintenance in industry, preventive medicine in human health, preventive detention of persons before social disturbances.
2. "Time, which changes people, does not alter the image we retained of them"- Ben Hect. This implies an important characteristic of information; an item of knowledge giving rise to information may change, but not the information itself.
3. "Time is a part of the framework of thought itself and without thought it cannot"- Christopher Fry. It shows the relationship between time and knowledge.
4. "Actually, time itself is neutral, it can be used either destructively or constructively"- Martin Luther King. It throws insight into the popular human trait attributing to time the good or bad

occurrences in one's life, rather than to one's own actions.

6. "Time is an invention of the west"- Uberto Eco. Reflects a truth perceived by the native Indian tribes of America. It also seems to hint the suppressed feeling of being a victim of White man's atrocities.
7. "Nine-tenth of wisdom consists in being wise in time"- Theodore Roosevelt. This quotation points out the importance of time in surmising a situation and in decision-making, i.e. about evaluation and use of knowledge/information appropriate to the occasion.
8. "Much of the wisdom of one age is folly to the next"- C. Simmons. Throws light on the validity of a given item of wisdom/knowledge over time.
9. "Rome is not built in a day." This proverb brings to mind the importance of cultivating patience in situations of over anxiety or utter frustration or anger. Patience being a time-delay mechanism helps in restoring balance of mind, in clear thinking, and in wise decision-making.
10. "Strike while the iron is hot"; "Make hay while the Sun shines." Implication of these sayings is that for successful outcome of any interaction one has to choose and act at the right time or critical time.

Time also has a psychological impact on human beings, which can bring solace, boost one's morale, act as a remedy, and change one's mood. This characteristic of time is reflected by the following quotes/proverbs: "Time eases all things"- Sopocles; "Time is the great physician"; "Time is a great healer".

5.5.2. Time-Related Sanskrit Sayings

"Kalaya tasmai namaha". This Sanskrit saying may be rendered in English as "Oh Time! Salute to you". It is paying obeisance to time for causing the happy ending of an event or fulfillment of a desire awaited for long. Hidden in the expression is also the sense of psychological relief. Because time is the controller of destiny and a great healer, enshrined in this saying is the Hindu religious and cultural ethos of reverencing all things and phenomena that are omnipresent and mysterious. Here, time is raised almost to the level of a deity as time is all pervasive and affects in some way or other everything in the cosmos, including human affairs. If this belief reflects the Hindus' sense of seeing divinity in all things and phenomena in nature, one could also argue that it represents the irrationality of mind in a scientific age where man is trying to control time through acquiring the ability to cause change or control the onset of change in many areas of life and nature.

5.5.3. Time-Related Kannada Proverbs

Atthegondhu kala, sosegondu kala : In English it means "bear in mind that a mother-in-law too had been once a daughter-in-law". It implies that the relationship between the two is to be based on understanding and not on intimidation and confrontation. This is particularly so in the context of Indian family-life where the relationship between mother-in-law and daughter-in-law is relatively more binding and intense than in certain other cultures. The Indian general perception, not always true, is that mothers-in-law are dominating over daughters-in-law.

Gidavagi baggaddhu maravagi baggithay : When as a plant doesn't bend, will it when it grows into a tree? The implied meaning is that it is easier to shape one's attitude/behaviour pattern in childhood or early stage of life rather than at the stage of adulthood.

5.5.4. Time-Related Tamil Proverbs

kARRuLLa pOtE tURRikkoL parallel in English 'Make hay while the sun shines'

tIyinAl cuTTa puN ularum ArATe nAvinAl cuTTa vaTu meaning The wounds of fire would vanish with time but the wounds caused by words never (Source: Thiruk-kural)

yAnaikkum aTi carukkum meaning Even the mighty slip at times.

5.6. CONCLUDING REMARKS

Life course perspective elaborates the importance of time,

context, process and meaning in human development and family life, (http://family.jrank.org/ pages/.....) and focuses on the relationship between individuals and the historical, socio-economic, and cultural context in which these individuals live(d), how the past and present contexts and events impact on individual's behaviour and may provide clues on the trends that may manifest in future. Thus, temporality, the concept of time is inherent in all aspects of life course. Social processes including family, education, employment, health, migration, etc. are affected by temporality. This paper briefly examined the relationship and role of time in selected segments of life course. We have also briefly reviewed studies and researches on the impact of time both from the philosophical and epistemological perspectives.

6

Time and the Arts and Humanities (with special reference to Indic Culture)

6.1. UNITY OF CONCEPTS AND TERMINOLOGY

A unity, a correspondence of concepts is discernible among the different art forms of Asia, such as, music, dance, sculpture, and architecture. For example, *taaLa* (*taaLam*), refers to measurement of "rhythm in music, cadence in dance, height in sculpture and area in architecture," in other words, *taaLa* is rhythm, measure and proportion for all the allied performing and visual arts. taaLa means palm of the hands. In architecture, the unit of measure is *hasta*, meaning hand, (in the West it is feet). This is evident in the floor plan of the Angkor Wat temple in Cambodia. A pair of cymbals tied together and struck with hands at intervals during chanting or a dance performance is also called *taaLa*. As the keeper of cosmic time, *taaLa* is an epithet of Siva, the Timeless. (Srinivasan, 2002, p.73). Srinivasan adds (p.75):

> *"In the medieval period, ragas (melodic modes) were ascribed to certain deities, in addition to each raga being*

associated with a specific colour, gender, mood, and so on. Therefore, the language of the sculptor, painter and musician adopts common terminology. A raga in music is also colour in visual art, masculine or feminine in quality, sentiments of attachment in philosophy, etc. Bhava is expression of mood in music, dance and sculpture as well. The integral nature of form and function in sound-space leads us from oral to aural to visual to emotional to spiritual experience in the seamless ocean of sacred arts whose interdependence was discovered long before the notion of multimedia began to shape our modern lifestyle.

"From silence to sound to syllable to word to poetry, thought flows into music, into order and dissolves into silence. The silent and active Brahman are not different ... Music has therefore has the extraordinary power to enable one to transcend sound and move into the mysterious sphere of silence; through time-bound rhythm cycles to timelessness, from sound space to Space. This is known as laya in Sanskrit." (Srinivasan, 2002, p.77).

6.2. TIME REPRESENTATION IN SPACE /ARTCHITECTURE

6.2.1. Hindu Religious Architecture

Hindu philosophy, religion and culture which had their origin in India later spread to several countries of Asia in particular. Many elements and aspects of the Indian cosmology, for instance, have been absorbed into the Indonesian Balinese dances, and the architecture of Cambodia's Angkor Wat temples, as in Indian classical dances and temple architecture. The expression of Time in Angkor Wat temple architecture is explained thus: "There were temporary wooden structures built for invoking the Spirit even in the pre-Vedic period, several centuries before the C.E. ..... The circular fire altar is made up of three hundred and sixty bricks to sympolise a revolution, a concrete representation of cosmic time. The imagery of the cosmic body or *Purusha* represented by the pole at the centre of the consecrated space-time altar called *Mahaavedi* corresponds to the human body with all its limbs, senses and apertures. This body in time establishes a relationship with space above and around, thus giving form to abstract speculative thought (Kak, 2000), quoted in Srinivasan, 2002). Further

> *"The scale of Angkor Wat is so expansive that it is difficult to comprehend both its structural and symbolic significance all at once. The entire sanctuary including its outer moat is based on Hindu cosmology of aeons or yugas. The duration of the earth is calculated as a mahayuga lasting 432 million years, which is a single day of Brahma, the Creator. At the end of it, the earth dissolves into the cosmic ocean for an equal period of time, regarded as the night of Brahma. The process of creation, preservation and destruction repeats thus endlessly in a cosmic cycle.*
>
> *"The aeons (yugas) are divided into four periods, namely Krta or Satya, Treta, Dvapara, and Kali. The duration of these four yugas are calculated to be in the ratio of 4:3:2:1, each of which precedes the other until the final cataclysm. In the Angkor Wat the distance from the bridge to the entrance across the moat is exactly 432 hastas or hands (used in measurement in temple building), marking the present aeon of Kaliyuga of 432000 years The distance between the outer wall entrance and the central enclosures marks the Dwaparayuga (864000 years), from enclosures to the western entrance of the central tower marks the Tretayuga (1296000 years) and from the central tower to the actual shrine the Krta or Satya yuga (1728000 years). The successively descending length of time as we come to the present Kaliyuga (the outer moat) is to suggest a corresponding physical and moral deterioration of people, the first yuga symbolically being*

the central tower, representing the golden period of creation.

"Even the measurements of the interior space of Angkor Wat, the number of towers, steps leading to the towers and the number of deities on the Sumeru coincide with the Solar and Lunar cycles. The image in the sanctum would have reminded the votary of that moment in cosmic creation when time and space merged at the centre and expanded to encompass the whole universe as Vishnu, in the form of Trivikrama. This compares with the Vedic ritual of Yajna that sought to transform the physical space-time to cosmic space-time in the belief that cosmic time could be restored to cosmic order through the holding of ritual." (Srinivasan, 2002, p. 12-13)

Time symbolism in the Chandi Borobudur architecture: "Gracing the entrances on all four directions are the awesome gates of of *Kaalamakara tOranas* of Yimein a monument of timelessness. For, it is the wheel of time that links beings in the heaven, earth or hell space in the relative universe of cause and effect" (Srinivasan, 2002, p. 24).

6.3. FOLK ARTS

Folk art is that form of art "not influenced by movements in academic or fine art circles, and, generally excludes works by professional artists, sold as "high art" or "fine art". Other overlapping terms, but with different connotations are: naïve art, Arts Primitive, Pop art, outsider art, traditional art, Tribal art, "tramp art", "self-taught" art and "working class" art. Folk art reflects traditional art forms of diverse community groups – ethnic, tribal, religious, occupational, geographical, age or gender-based – who identify with each other and society at large. Folk artists traditionally learn skills and techniques through apprenticeship in informal community contexts. Several forms of classical music, drama, contemporary paintings etc., have arisen out of or have drawn elements from folk arts.

India has a rich repertoire of folk and tribal arts with variations and distinctiveness in terms of time from one region to another. Some of the more well known Folk and Tribal Art of India are:

- Tanjore Art;
- Madhubani Painting;
- Warli Folk Painting;
- Pattachitra Painting;
- Rajasthani Miniature Painting; and
- Kalamezhuthu

(*Source*: National Portal Content Management Team Reviewed on: 31-08-2010)

Folk art expresses cultural identity by conveying shared community values and aesthetics. It covers a range of utilitarian and decorative media, including cloth, wood, paper, clay, metal and other media as in contemporary expressions of traditional folk art forms.

The 35 states and union territories sprawling across India have their own distinct cultural and traditional identities, and are displayed through various forms of art prevalent there. Every region in India has its own style and pattern of art, which is known as folk art. Other than folk art, there is yet another form of traditional art practiced by several tribes or rural population, which is classified as tribal art. The folk and tribal arts of India are very ethnic and simple, and yet colorful and vibrant enough to speak volumes about the country's rich heritage. Many volumes can be filled in describing the folk arts of India.

Several generations of tribal communities (over 32) of the State, since time immemorial have conceived and shaped its culture and arts. "Cross-cultural influences of local non-tribal communities and successive waves of Buddhism and Jainism, Mughal rule and the reign of the Hindu emperors of Bengal."

Old cave paintings and Stone Age tools discovered in Hazaribagh district and axes and spearheads found in the Chaibasa area, are remnants of a civilization dating back a few thousand years.

10,000 to 30,000 years old rock paintings, paintings in huge caves in the Sati hills and other indicators of ancient, even pre-historic, human settlements of the region.

These folk art forms and themes are very closely associated with nature and its offerings. The performances are also timed to coincide with particular seasons, festivities and functions (e.g. welcoming guests). Karma, for example, is a popular folk dance performed in the month of August during the Karma festival. The following brief note on the folk arts of Jharkhand State, North India, is just an example:

Art and Craft of Jarkhand–Timing : Archaeologists have unearthed pre-Harappan pottery, and pre-historic cave paintings and rock-art that hint at ancient, cultured civilizations inhabiting these parts. The intricate woodwork, the pitkar paintings, tribal ornaments, stone carvings, dolls and figurines, masks and baskets, indicate how deep into time these manifestations of culture go, how the well-spring of creativity continues to recharge the spirit of the tribes and the state itself.

The Kohvar and Sohrai paintings are among the most delicate, fragile, beautiful and almost disappearing indigenous traditions of India. These are sacred, secular and relevant and practised exclusively by married women, during weddings and at harvest time, and the skill and the information is handed down to younger females of the clan.

6.4. INDIAN MUSIC: TIME-THEORY BASED RAGA CLASSIFICATION

Time in Indian Music by Clayton (2008) is a study of rhythm, metre, and form in North Indian *rag*, or classical music. He presents a "theoretical model for the organization of time in this repertory, a model which is related explicitly to other spheres of Indian thought and culture as well as to current ideas on musical time in alternative repertories including that of Western music." The examples given help to clarify key Indian musicological concepts such as *taal* or *taala* (metre), *laya* (tempo or rhythm), and *laykari* (rhythmic variation).

The work addresses the implications of performance practice for the organization of rhythm and metre.

A highly developed system of music had evolved in early times in most parts of India. The Tamils (in South India) developed a music system of their own.... The Tamils had an advanced system of music and dance even before the time of Bharatha, that is between 200 B.C.E. and 1200 A.D. Many of the Sangam works refer to music prevalent in the Tamil region. The period of the Saiva saints and their contribution to Tamil music is quite significant. In Tamil, music is denoted by the term *isai.* The division of Tamil land into different regions, like pastoral region (*mullai*), arable land (*marutham*), hilly areas (*kurinjci*), waste land (*palai*), coastal region (*neytal*), etc., became a basis for classification of ragas. They developed *pann*' of their own and had different systems of music.

Carnatic music represents the rich cultural heritage of South India and the essence of spirituality that evolved out of the heart of pious spiritual gurus of the past. The origin of Carnatic music can be traced back to the Vedas. Bharata's *Natya Sastra* (c.5th century CE) and Saranga Deva's *Sangita Ratnakara* (early 13th century CE) are the earliest documents on the theory and performance of Indian classical music. Carnatic music has as its basis a system of *ragas* (melodic time scales) and *talas* (rhythmic cycles). The rhythmic cycles are 7 in number and there are 72 fundamental ragas called *melakarta ragas*, a graphical presentation of which is given in Annex (Neelameghan and Raghavan, 2012).

Time for Music : The *panns* were classified on the basis of the time theory – *pakalpanns* and *iravupanns.* For *potupanns* the classification was based on *ganakala.* These ragas were intended to be sung during different times of the day. Of the 24 *panns,* 12 were *pakalpanns* (day ragas; 4 AM-10 AM), 9 *iravupanns* (night ragas) and 3 *potupanns* (sarvakalika ragas). The 12 *pakalpanns* are: *nattappadi, puraniramai, gandharam, piyandaigandharam, kausikam, indoLam, takkesi, sadari, mattaragam, pazhampanjuram, gandhara panchamam,* and *panchamam.* The ragas *bhupaLam, bowli, revagupti* are considered particularly suited for early morning, 4 to 6 AM., and the ragas bilahari, malayamarutham and saveri are suitable for the morning period 6-10 AM.; The ragas *begata, danyasi, madhyamavati, bhairavi* are deemed particularly suitable for the period 10 AM to 4 PM. (Pamchapakesa Iyer, 1989). The *iravupanns* are: *takka ragam, pazhamtakka ragam, sikamaram, kolli, kollikauvanam, viyazhakkurinji, megharagakkurinji, andalikkurinji*, and *kurinji*. According to him for the period 4 PM to 8 PM, ragas like *sankharabharanam, kalyani, kambhoji, bhairavi, thodi, kharaharapriya* and *pantuvarali* are suitable. He considers ragas like *nilaambari, ananda bairavi* and *edukula kambhoji* suitable as night ragas (8 PM onwards) The *potupanns* are: *senturutti, thiruthandagam* and *sevvazhi*.

Dutt (2012) reporting about a 'Chaturprahar' presentation by noted Indian musicians wrote: "According to Indian classical music tradition, ragas can be experienced in all their splendor only when played at a designated time. It is believed that there is a connection between the dominant mood that a raga invokes and the bodily and psychological changes that occur within us as the day progresses and time zones change..... this has to do with the evolution of music over centuries when musicians and their listeners lived in proximity to the natural world, its sights, sounds and rhythms."

The Chaturprahar divides the dusk to dawn period of 12 hours (6 p.m. to 6 a.m.) into 4 time zones of 3 hours each. "At dusk, or 6 p.m. as the sun begins to set, the mood is ambiguous and a strange restlessness surfaces. This mood finds resonance in the ragas *marwah or shree, durga* or *bahar*. During the next time zone (9 p.m. to midnight) "shades of love, devotion and surrender overlap to create a mood of longing." This mood is said to resonate with the ragas *yaman kalyan, shud kalyan .... khamaj thaat raga, jhinjothi* or *jaijaijayawanti*. The raga *paramelapraweshak*from *kafi thaat*marked the transition from the early to the dense part of the night. Next, redition in *dharbari kannada* with detailed *alaap* and *dhrupad* in *chautal* marks the stillness of the deep night; this ends with short renditions in *Shiddha Kauns*and *Sohini*. The closing renditions marked the early dawn with presentation of a rare raga *Lalit* followed by short early morning ragas concluding with *bhairavi.*

taaLa or taal : In Indian classical music *tala* literally meaning a 'clap', referring to the "rhythmic pattern of any composition, roughly corresponding to metre in Western music, though closer conceptual equivalents are to be found in other Asian classical systems such as the notion of *usul* in the theory of Ottoman/Turkish music." (Wikipedia). *taaLa* is based on a cycle with the centre space s*ama.* In comparison to a pulse which merely beats in a linear fashion without a centre, rhythm in the Indian context has a centre and its symmetry, giving the cyclic pattern .... Starting with a 3 beat cycle, Indian rhythm can comprise of any number of combinations of 4,6,7,10,12,16,64 or 108 beat cycles, but each must end where it begins, the centre. ... The percussion provides the metrical framework for the melody and also complements and heightens any improvisations by the musician. "If we take the human voice as a universal musical instrument, the attributes of voice is determined by what the voice is expected to do. This becomes culture-specific (Srinivasan, 2002, p.74).

Rhythm in Indian music performs the function of a time counter. A *taal* is a rhythmic cycle of beats, as already mentioned with an ebb and flow of various types of intonations resounded on a percussive instrument. Each such pattern has its own name. Indian classical music has complex, all-embracing rules for the elaboration of possible patterns, though in practice a few *taals* are very common while others are rare. The most common *taal* in Hindustani classical music is *teental,* a cycle of four measures of four beats each.

A *taal* does not have a fixed tempo and can be played at different speeds. In Hindustani classical music a typical recital of a raga falls into two or three parts categorized by the tempo of the music - *vilambit laya* (Slow tempo), *madhya laya* (Medium tempo) and *drut laya* (Fast tempo). In Carnatic music, there are five categories of tempo namely - *Chauka* (1 stroke per beat), *vilamba* (2 strokes per beat), *madhyama* (4 strokes per beat), *dhuridha* (8 strokes per beat), *Adi-dhuridha* (16 strokes per beat). Although the tempo changes, the fundamental rhythm does not. Each repeated cycle of a *taal* is called an *avartan.* A *tala* is generally divided into sections (*vibhaags*), not all of which may have the same number of beats.

In a survey of Carnatic music P. Sambamoorthy (1983; 1987) describes certain aspects of the *tala* system: "... The *sula sapta talas* give rise to 35 varieties on account of the 6 varieties of *laghu*. These again give rise to 175 varieties on account of the *pancha gati bheta*. There is also another set of 5 *laghus*: *divya laghu* (6 units of time), *simha laghu* (8 units of time), *varna laghu* (10 units of time), *Vadhya laghu* (12 units of time) and *Karnataka laghu* (16 units of time). This when applied to the *suladi sapta talas* will give rise to another 35 *talas*, which with the *pancha gati-bheda* will give rise to another possible 175 *talas*. Besides these there are *sankirna talas* or rhythms of mixed pattern used in scholarly *pallavis* and in the folk music of the land".

The most common instrument for keeping rhythm in Hindustani music is the *tabla*, while in Carnatic music, it is the *mridangam*.

Western music may be categorized by time periods as shown in the following table (From Wikipedia, the free encyclopedia):

Periods of European art music

Early	
Medieval	(500–1400)
Renaissance	(1400–1600)
Baroque	(1600–1760)
Common practice	
Baroque	(1600–1760)
Classical	(1750–1830)
Romantic	(1815–1910)
Modern and contemporary	
20th-century	(1900–2000)
Contemporary	(1975–present)
21st-century	(2000–present)

The dates of the Classical Period in Western music are generally accepted as being between 1750 and 1830. However, the term *classical music* is also popularly used to describe a variety of Western musical styles from the ninth century to the present, and especially from the sixteenth or seventeenth to the nineteenth. The Classical period falls between the Baroque and the Romantic periods.

6.5. DANCE

Western dance is concerned more with space, and the dance movements involve moving around in space. Hence they are characterized by great leaps into the air, or the rapid movement from one place to another. Western dancers create movement by reaching into space and carving out intricate movements in the air, and the central aim is to make movement appear free from gravity. Indian dancers on the other hand have a very different concept they do not concern themselves with spaces but are concerned with time. The central aim of Indian dance is to make a pose so perfect that it appears timeless. Hence the movements take place in time and must be perfectly synchronized with the rhythm, in order to create a series of perfect poses. A perfect pose would be able to capture time for a moment, and in a limited space. As a result most Indian dances, except for certain movements in *Kathakhali*, do not feature any major leaps or other sudden changes in space. (Wikipedia)

The principal function of the sacred dances of Bali, Indonesia, is the "effacement of individual self and submerging of one's own identity in the eternal flow of movement." The Balinese dance may be viewed from two perspectives: "the time-specific or *sekala* and *nishkala* or timeless, with deep symbolic implications." Radhika Srinivasan explains the implications by elaborating on the Barong-ket theatrical dance tradition of Indonesia – the encounter between good (represented by *dewa Ayu*) and evil (represented by *rangda*), and eventual triumph of the former and defeat of the latter.

> *"... Through time subdivided into moments, the dancer's rhythm seeks to arrest the votary's sense of time in order to savour a brief sense of timelessness..." (Srinivasan, 2002, p.52)*

In the ancient Indian classical dance text *Natyasastra* authored by Bharata it is said: from the half-seated basic posture, the dancer executes geometric patterns of 108 *mandalas* through cadences that give primacy to a time cycle that is divided into beats of three, four, six, eight, ten or sixteen. Six different postures (*sthana*) are described. ... "The entire performance becomes a process through which cosmic Creation, Preservation and Destruction are replicated on stage. A classical Bharathanatyam repertoire typically begins with an invocatory item based on unintelligible sound syllables uttered by the guru and the percussionist. The precedence of percussion to all other musical instruments and to the rhythmic cycle of five beats is intended to invoke the elements is like paying obeisance to movement itself, This item is deliberately devoid of lyrics because the Vedas describe the evolution of life to have begun with unintelligible sound bursting out of silence at the time of genesis.

Measuring Beats : The metronome is a tool for producing steady beat (pulse) enabling musicians play rhythms correctly, keeping time and counting beats. The pulses are measured in beats per minute (BPM). The metronome can play beats from about 35 to 250 BPM.

Folk Dances – India : Folk dance and music are not just for entertainment. They are the means to invoke / propitiate a chosen deity during certain seasons or times and / or for certain purposes, e.g. propitiate the rain god, ward of evils, natural disasters foreseen, and to signify the time of arrival of seasons, the birth of a child, weddings, festivals, etc.

Living Rhythms : Every ethnic community has the tribal memory of its culture and arts. The communities' tradition of oral transmission of knowledge and experience over the centuries, substantial parts of this memory bank have been lost. However, in recent years there has been a revival. For instance, "the Adivasis of Jharkhand are coming back, to the strong sounds of drums, flutes, cymbals and voices raised in song. When the tribes of Jharkhand gather to celebrate a very special occasion, whether at home or in village society, they make music and dance to its beat / rhythm. Perhaps it makes them happy because it brings that long memory back, into the framework of real life. They yearn to carry ancient memory and culture into their lives today.

Jharkhand tribes use many kinds of percussion, along with a variety of musical instruments to accompany their singing and dancing. The Nagara, made of wood and stretched animal hide played with drumsticks sound the best in summer, in winter they lose their vibrant beat.

Dance forms echo warrior-like movements of battles fought long ago, for the men. Observations of animal and bird behaviour were also blended into the dancing, sometimes light and airy, sometimes heavy and solemn, as the mood demands. Everyday activities of womenfolk - working in the fields, in the forests, in their homes, all reflect themselves in dance forms like the timeless Chhau, which requires agility, suppleness, skilful hand movements. The Seraikela school of Chhau is held during the Sun Festival. It is influenced by the principles of Bharatiya Natya Shastra but the theme is based on folk-lore, nature and mythology. The choreograph follows Odissi in composition and execution.

Other well-known dances of tribal communities in Jharkhand include festive dancing such as - Sarhul / Baha, where sal and mahua flowers are used ceremonially; the Dansai amd Sorhai where the youth sing and dance through the night of karam; Maghi Puja, an important festival of the Munda tribe; Sarhul in which 'shalai' the flower of the sal tree is offered to the gods, with the flower as a symbol of brotherhood; Tusu, the harvest festival, celebrated mainly by unmarried girls."

6.6. PSYCHIC ABILITIES (PSI) AND TIME

Psychic abilities displayed by some people include extrasensory perception (ESP), telepathy, precognition, telekinesis, clairvoyance and related ways of acquisition of information and human communications. In India and some other eastern countries such phenomena are not uncommon and generally accepted. In the West, although there have been several studies in this domain during the last couple of centuries, it was only with the arrival of J.B. Rhine at Duke University and the establishment of the Parapsychology Laboratory in 1927, the researches carried out and papers published gave credence to parapsychology as an experimental science. Rhine's book *Extra-Sensory Perception* (1973/74) and *Extrasensory Perception after Sixty Years* by Rhine and others recording 145 experimental studies of ESP further promoted serious studies by researchers at this interesting field of parapsychology at several other centres in the West. The field "has crossed many milestones, overcoming many hurdles and accumulated massive evidence..." (Ramakrishna Rao, 2010)

Research in parapsychology "has gone well beyond the question of existence of psi. There has been experimental demonstration of various forms of psi and the discovery of lawful regularities between psi and other psychological and physical variables. *One of the most important finding is that ESP, unlike any known physical phenomenon, is apparently unconstrained by space/time variables or the complexity of the task* (Stanford, 1977). An example from Srinivasa Ramanujan's biography by S.R.Ranganathan: (Ranganathan, 1967)

> *"Posthumous Phenomenon:*
>
> *I may here refer to an even more uncanny occult happening. It was 1934. I was with S. Krishnaswami Ayyangar, a leading advocate of Madras and later a judge of the High Court. He had known Ramanujan as a boy in Kumbakonam. With the help of the Ouija board, he invoked Ramanujan. There was response. I asked him whether he was continuing his mathematics. He replied, 'No. All interest in Mathematics dropped out after crossing over.' I asked him what he was occupying himself with. The reply was, Meditation. Studying Vishnu-sahasrananama (a hymn of 1,008 names of God). I then told him about his third Notebook which was in my possession in the University Library. Only a few pages of it had been filled up. It was all a series of tabulated numbers. Nobody could make out what it was about. Could he throw any light on it? Ramanujan replied, 'I do not remember about it. Bring the Notebook to this place this time next week. I shall look into it. If I*

can recollect anything, I shall tell you.' The Notebook was taken to Krishnaswamy Ayyangar's house st the appointed time. Ramanujan was invoked. He asked me to turn the pages slowly. Then came the reply, 'I now remember. I was working on Mock-Theta Function.' This brought to my mind Prof G.H.Hardy's reference to Ramanujan starting work on the Mock-Theta Fnction, before he left Cambridge in 1919."

6.7. LITERATURE

As in other fields of study the concept of system / style of different periods exists in literature (*belle lettres*) (prose, poetry, drama, etc.). Example: primitive / ancient, classical, medieval, Renaissance, Elizabethan, modern, post-modern, etc. In documentary classification schemes such concepts are usually represented by the period (century, years, decades) when the styles originated or were prevalent.

In Colon Classification (CC) the facets for 'Literature' are Language, Form, Author, and Work. The objective is to bring together documents (and their surrogates in the classified part of a Classified catalogue) by language, within a language group by form of work (poetry, prose, drama, etc.), within each form division (group), by the 'Author' and within this by the work of each author. The evaluations and commentaries for each work will get arranged together. The 'Author' isolate number is derived by the Chronological Device – the date of birth of the author (century, decade, month) as available. For details and guidelines see Ranganathan's Colon Classification.

6.8. CLASSICS

A classic is a work of the highest excellence. "A book stimulating other books and literature on itself. The inherent qualities of a classic work include:

1. It has elements of permanent value;
2. It is saturated with the personality of the author itself was very powerful and highly organized; and

3. It is a seminal work cutting new ground, blazing new trail, stimulating new thought, and so on. (Ranganathan, 1960)

The author of a classic work is deemed a Classical Author. There are classic works in almost all subject fields; they are more prominent and numerous in the humanities.

CC uses the "Classic Device (ClD) for bringing together the different editions of a classic in a class, the different editions of each of its commentaries, the different editions of each of the sub-commentaries of a commentary, and so on, and of securing that the group of sub-commentaries are in juxtaposition to the classic, and the group formed of each classic and the associated commentaries is in juxtaposition to the groups of the other classics of the same class."

"The CID consists in putting the digit x after the number representing the ultimate class to which the classic should otherwise be assigned and adding after that an Author Facet and also a Work Facet .... As in the case of the class number construction for 'Literature' See Sec. 7". For details and guideline refer to Ranganathan, 1960, chapter 7. CC edition 6 Part 3 provides an extensive schedule of classics (of India in particular) in various subjects and for sacred books treated as classics.

6.9. TIME, SPACE AND CONTEXT IN TEXT SCIENCE

Text science aims at establishing universal rules applicable to all forms of human expression and communication. 'Text' includes any meaningful expression – written, oral discourse and dialogue, figures, images, icons and symbols, gestures and body actions, etc. "Text science attempts to explain how these rules function throughout various forms of text and thus provide a better understanding of human behavior." A particular guideline / rule is that in evaluating a text in different languages, and in different editions) or commentaries thereof, the author of the latter should consider the context - the time and place and the social environment of the author of the original text being evaluated / commented upon. (Wada, 2010, Shimoda, 2010)).

Ganesh (2011) presents a parallel view in his analytical study of the Tamil grammar classic *Tolkappiyam* pointing out that *Tolkappiyam* (c.3rd century BCE – 10th century CE) is not just about Tamil linguistics and Tamil grammar but an important source of information about Tamil society at that time.

> *"Several studies on the early Tamil texts have appeared from linguistic, cultural and historical perspectives, which have helped to provide valuable insights into structural and procedural aspects of early Tamil society. Studies of Tinai and Turai have been central to these efforts. .... Tolkappyiam is primarily a text on Tamil language and grammar, but its significance in understanding the society of early Tamilakam has long been recognized. It is also known that Porulatikaram in Tolkappiyam goes far beyond the conventional linguistic treatises in dealing with social processes...... it provides a comprehensive methodology in understanding early Tamil texts. ... the methodology involves historical categories of space and time and formation of events that constitute the basis of literary expression. It is argued that the context of introduction of these categories (space-time) signifies major transformation in early Tamil society, which makes the composition of Tolkappiyam representative of changing social order. (Ganesh, 2011, abstract)*

ANNEXURE

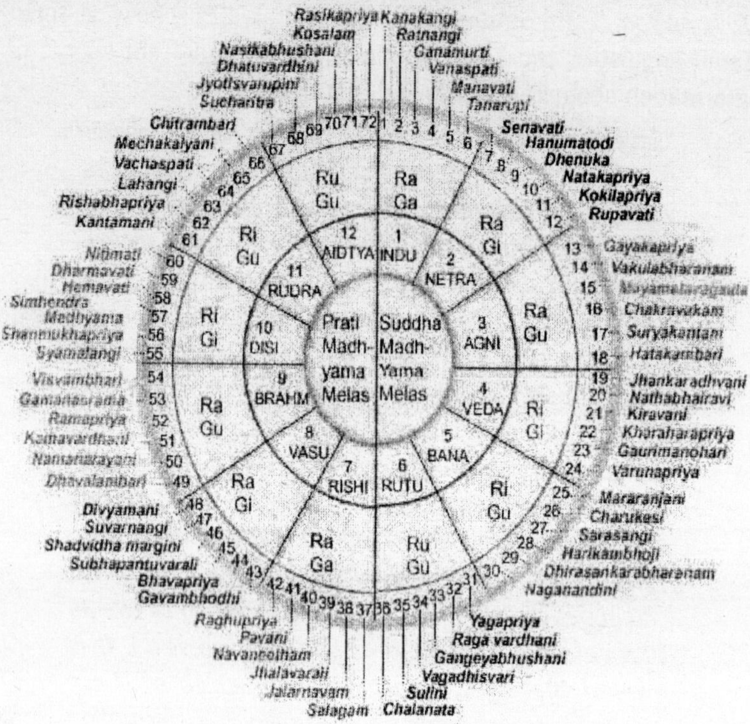

Fig. 1. The Melakarta ragas of Carnatic Music

(*Source* : www.carnaticindia.com)

An alternative categorization and tabular presentation of the Melakarta ragas is given below. (*Source: http://en.wikipedia.org/wiki/Melakarta#Table_of_Melakarta_ragas*)

'Melakartâ Râgas'

'Shuddha Madhyama'

No.	Raga	Scale
		1. Indu Chakra
1	Kanakangi	$S\ R_1\ G_1\ M_1\ P\ D_1\ N_1\ S'$
2	Ratnangi	$S\ R_1\ G_1\ M_1\ P\ D_1\ N_2\ S'$
3	Ganamurti	$S\ R_1\ G_1\ M_1\ P\ D_1\ N_3\ S'$
4	Vanaspati	$S\ R_1\ G_1\ M_1\ P\ D_2\ N_2\ S'$
5	Manavati	$S\ R_1\ G_1\ M_1\ P\ D_2\ N_3\ S'$
6	Tanarupi	$S\ R_1\ G_1\ M_1\ P\ D_3\ N_3\ S'$
		2. Netra Chakra
7	Senavati	$S\ R_1\ G_2\ M_1\ P\ D_1\ N_1\ S'$
8	Hanumatodi	$S\ R_1\ G_2\ M_1\ P\ D_1\ N_2\ S'$
9	Dhenuka	$S\ R_1\ G_2\ M_1\ P\ D_1\ N_3\ S'$
10	Natakapriya	$S\ R_1\ G_2\ M_1\ P\ D_2\ N_2\ S'$
11	Kokilapriya	$S\ R_1\ G_2\ M_1\ P\ D_2\ N_3\ S'$
12	Rupavati	$S\ R_1\ G_2\ M_1\ P\ D_3\ N_3\ S'$
		3. Agni Chakra
13	Gayakapriya	$S\ R_1\ G_3\ M_1\ P\ D_1\ N_1\ S'$
14	Vakulabharanam	$S\ R_1\ G_3\ M_1\ P\ D_1\ N_2\ S'$
15	Mayamalavagowla	$S\ R_1\ G_3\ M_1\ P\ D_1\ N_3\ S'$
16	Chakravakam	$S\ R_1\ G_3\ M_1\ P\ D_2\ N_2\ S'$
17	Suryakantam	$S\ R_1\ G_3\ M_1\ P\ D_2\ N_3\ S'$
18	Hatakambari	$S\ R_1\ G_3\ M_1\ P\ D_3\ N_3\ S'$
		4. Veda Chakra
19	Jhankaradhwani	$S\ R_2\ G_2\ M_1\ P\ D_1\ N_1\ S'$
20	Natabhairavi	$S\ R_2\ G_2\ M_1\ P\ D_1\ N_2\ S'$
21	Keeravani	$S\ R_2\ G_2\ M_1\ P\ D_1\ N_3\ S'$
22	Kharaharapriya	$S\ R_2\ G_2\ M_1\ P\ D_2\ N_2\ S'$
23	Gourimanohari	$S\ R_2\ G_2\ M_1\ P\ D_2\ N_3\ S'$

24	Varunapriya	S R_2 G_2 M_1 P D_3 N_3 S'

5. Bana Chakra

25	Mararanjani	S R_2 G_3 M_1 P D_1 N_1 S'
26	Charukesi	S R_2 G_3 M_1 P D_1 N_2 S'
27	Sarasangi	S R_2 G_3 M_1 P D_1 N_3 S'
28	Harikambhoji	S R_2 G_3 M_1 P D_2 N_2 S'
29	Dheerasankarabharanam	S R_2 G_3 M_1 P D_2 N_3 S'
30	Naganandini	S R_2 G_3 M_1 P D_3 N_3 S'

6. Rutu Chakra

31	Yagapriya	S R_3 G_3 M_1 P D_1 N_1 S'
32	Ragavardhini	S R_3 G_3 M_1 P D_1 N_2 S'
33	Gangeyabhushani	S R_3 G_3 M_1 P D_1 N_3 S'
34	Vagadheeswari	S R_3 G_3 M_1 P D_2 N_2 S'
35	Shulini	S R_3 G_3 M_1 P D_2 N_3 S'
36	Chalanata	S R_3 G_3 M_1 P D_3 N_3 S'

'Prati Madhyama'

7. Rishi Chakra

37	Salagam	S R_1 G_1 M_2 P D_1 N_1 S'
38	Jalarnavam	S R_1 G_1 M_2 P D_1 N_2 S'
39	Jhalavarali	S R_1 G_1 M_2 P D_1 N_3 S'
40	Navaneetam	S R_1 G_1 M_2 P D_2 N_2 S'
41	Pavani	S R_1 G_1 M_2 P D_2 N_3 S'
42	Raghupriya	S R_1 G_1 M_2 P D_3 N_3 S'

8. Vasu Chakra

43	Gavambhodi	S R_1 G_2 M_2 P D_1 N_1 S'
44	Bhavapriya	S R_1 G_2 M_2 P D_1 N_2 S'
45	Shubhapantuvarali	S R_1 G_2 M_2 P D_1 N_3 S'
46	Shadvidamargini	S R_1 G_2 M_2 P D_2 N_2 S'
47	Suvarnangi	S R_1 G_2 M_2 P D_2 N_3 S'
48	Divyamani	S R_1 G_2 M_2 P D_3 N_3 S'

9. Brahma Chakra

49	Dhavalambari	S R_1 G_3 M_2 P D_1 N_1 S'
50	Namanarayani	S R_1 G_3 M_2 P D_1 N_2 S'
51	Kamavardani	S R_1 G_3 M_2 P D_1 N_3 S'

52	Ramapriya	$S\ R_1\ G_3\ M_2\ P\ D_2\ N_2\ S'$
53	Gamanashrama	$S\ R_1\ G_3\ M_2\ P\ D_2\ N_3\ S'$
54	Vishwambari	$S\ R_1\ G_3\ M_2\ P\ D_3\ N_3\ S'$
	10. Disi Chakra	
55	Shamalangi	$S\ R_2\ G_2\ M_2\ P\ D_1\ N_1\ S'$
56	Shanmukhapriya	$S\ R_2\ G_2\ M_2\ P\ D_1\ N_2\ S'$
57	Simhendramadhyamam	$S\ R_2\ G_2\ M_2\ P\ D_1\ N_3\ S'$
58	Hemavati	$S\ R_2\ G_2\ M_2\ P\ D_2\ N_2\ S'$
59	Dharmavati	$S\ R_2\ G_2\ M_2\ P\ D_2\ N_3\ S'$
60	Neetimati	$S\ R_2\ G_2\ M_2\ P\ D_3\ N_3\ S'$
	11. Rudra Chakra	
61	Kantamani	$S\ R_2\ G_3\ M_2\ P\ D_1\ N_1\ S'$
62	Rishabhapriya	$S\ R_2\ G_3\ M_2\ P\ D_1\ N_2\ S'$
63	Latangi	$S\ R_2\ G_3\ M_2\ P\ D_1\ N_3\ S'$
64	Vachaspati	$S\ R_2\ G_3\ M_2\ P\ D_2\ N_2\ S'$
65	Mechakalyani	$S\ R_2\ G_3\ M_2\ P\ D_2\ N_3\ S'$
66	Chitrambari	$S\ R_2\ G_3\ M_2\ P\ D_3\ N_3\ S'$
	12. Aditya Chakra	
67	Sucharitra	$S\ R_3\ G_3\ M_2\ P\ D_1\ N_1\ S'$
68	Jyoti swarupini	$S\ R_3\ G_3\ M_2\ P\ D_1\ N_2\ S'$
69	Dhatuvardani	$S\ R_3\ G_3\ M_2\ P\ D_1\ N_3\ S'$
70	Nasikabhushani	$S\ R_3\ G_3\ M_2\ P\ D_2\ N_2\ S'$
71	Kosalam	$S\ R_3\ G_3\ M_2\ P\ D_2\ N_3\ S'$
72	Rasikapriya	$S\ R_3\ G_3\ M_2\ P\ D_3\ N_3\ S'$

7

Time in Information Science

7.1. NOTION AND DEFINITION OF INFORMATION

To understand the temporal aspects of time in information science, it is helpful to begin with 'information.' The notion of information is better explained by beginning with 'natural' information. Natural information is the pattern of organization of matter and energy in the material world. Natural information gets represented through the process of encoding or embedding. The encoded information, which we encounter in our social and perceptual world, is based on patterns of organization derived from the use of symbols, language, and signals. Embedded information is the corporeal expression or manifestation of information previously encoded. Thus, information "given meaning and integrated with other contents of understanding" is knowledge (Bates, 2006, p.1044). Taking a reverse view, information appears as the external manifestation of knowledge in which form it is named 'explicit' or embodied knowledge. Hence, the study of time-space effects on knowledge is also relevant to information science.

The definition of information science and its contents is changing with time. This is clearly reflected by the definition adopted by *Information Science Abstracts* (ISA) of USA. ISA defines information science as "an interdisciplinary field concerned with the theoretical and practical concepts as well as the technologies, laws, and industry dealing with knowledge transfer and the sources, generation, organization, representation, processing, distribution, communication and uses of information, and communications among users and their behaviour as they seek to satisfy their information needs" (Hawkins, 2001, p.49). There is a major shift of emphasis in this definition from library science / documentation bias to that of the Internet industry dominated world. The shift of emphasis is accentuated by the dynamic changing situation.

Any definition of information science to be valid today has to accept and reflect upon the nature, form and complexities of information and information systems existing in the digital era. This change is a natural consequence of the application of Ranganathan's fourth and fifth laws of library science (as adapted to the context of information science and information systems), namely, "Save the time of the reader / Save the time of the staff", and "Library is a growing organism" respectively (Ranganathan, 1931). The impact of these laws is directly related to economizing time on the part of the library / information user and of the library staff and to the effects of change, growth (physical as well new resources, techniques etc., replacing outdated ones). Change occurs in a time / space matrix, and has a reciprocal relationship with knowledge.

7.2. PRODUCTION AND QUANTIFICATION OF INFORMATION

The world in which we are living today is an information world, which is dynamic and complex. The amount of information is increasing day by day, if not by the hour. Quantifying the amount of information that exists in the world is difficult. It is estimated that it is growing at a compounded annual rate of 60%. Digital information according to an IBM study, cited by Martin Limonica of news.com, doubled every 11 hours in 2010 (Data, data, 2010). "The flood of data from sensors, computers, research labs, cameras, photos, phones and the like surpassed the capacity of storage technologies in 2007"(Data, data, 2010). It is no more a world in which information is structured, homogenous, with emphasis laid on information leading to knowledge and scholarship. The 2010 special report on managing information by *The Economist* indicates that only 5% of the information that is created is 'structured', meaning that it comes in a standard format of words or numbers that can be read by computers. The rest consists of items like photos, phone calls, data collected from social and commercial transactions conducted over Internet, social networks and websites. There is more of real-time data, their analysis and utilization than the one based on 'dead' data in the world of commerce including economics of Internet and websites.

"The Digital Universe in India"- a report issued under the sponsorship of EMC, an information infrastructure technology firm, projects that the digital information in India will grow from the present 40,000 petabytes to 2.3 million petabytes by 2020 - a growth of 60 times in just 10 years. This is accounted for by the launch of 3G and BWA networks, digitization of TV networks, increased technology use by individuals, small and medium size enterprises, census and Unique ID project initiated by the Government of India (Chugh, 2010).

7.3. COMPRESSIBILITY AND EXPANDABILITY: INFORMATION AND TIME

7.3.1. Compression

Brevity of expression is a desirable characteristic for effective human communication and compression of information is the surest way of achieving brevity.

In representing information for presentation and later retrieval, brevity of expression plays a vital role. Human mind prefers to receive and act quickly upon input information if the content is presented in a compact form. The compactness in expressing the contents of a scientific text is achieved by compressing information, apart from the use of scientific terminology. In other domains, with the possible exception of literature, brevity is achieved by the process of selection and use of precise and unambiguous terminology to represent the contents and their interrelations in a text in the shortest form possible. Brevity and compactness make remembrance easier, comprehend the essence of a text quicker; economize time and neural energy. From the point of view of time, condensing of information saves time of the user of the text by indicating whether the user should go for the full text or limit to the brief. It also saves communication time. Some of the means of accomplishing brevity are mentioned below.

In scientific communication, compression of data is of critical importance in representing information in a compact form. Information is compressed by identifying and using structures that exist in the data. Compressibility is a property of information that links closely information with physics, and indeed with all of science (Feynman, 1963). Deployment of brevity of expression and the use of command are the two most important means of imparting compressibility to information. Brevity retains cognitive links, which is an absolute requirement, and command works as a mechanical device. In science and mathematics compressibility is achieved by the use of unique symbols/notation, framing of laws/axioms, use of equations and formulae, as could be seen from the two examples cited by Von Beayer (2004).

1. Newton's laws of motion and gravity. These express in a compressed form the information about the orbits of all the planets known or yet to be discovered, and also the motions of their parent stars, as well as those of footballs, tidal waves and other massive bodies in the universe.
2. Einstein's relativistic field equation $R_{\mu\nu} - \frac{1}{2} g_{\mu\nu} + g_{\mu\nu} = 8\pi G T_{\mu\nu}$. The equation condensed all the information contained in the original paper — fifty pages devoted to explaining the symbols used, sixteen interlinked equations and several hundred entries needed to define a position and an instant in time-space besides a very large number of relationships, descriptions of geometry of space-time, and distribution of sources of gravity covered by the equation. The amount of information in this one equation is so overwhelming that nobody can make use of it in full generality.

To give an example from the domain of classification (library science) is the formula (P);(M):(E).(S)'(T), which contains the entire philosophic base and the structure of faceted classification scheme, such as the Colon Classification developed by S.R. Ranganathan, in a very compressed form.

A more common example for compression of information is a picture, whose impact is succinctly conveyed by the proverb "A picture is worth a thousand words". The greatest device that compresses, as well as expands information is the human brain. Other devises invented to compress information include symbols and formulae in chemistry, equations in science and mathematics, and sutras in Sanskrit grammar.

Physical compression of information is achieved by the use of symbols as is the case in a language and in chemistry; and by electrical charges (coded as 0 and 1) and bits and bytes in digital communication, though in the latter case compression is only an intermediary stage involving physical transmission and storage of information in a coded form. Without compression of data it would be near impossible to have Internet, cellular phone or digital TV (Sayood, 2000). Transmitting of coded information by binary digits 0101010101……. in the compressed form achieved by a single instruction "Repeat 01 without stopping" is an example of compressibility by command.

In transmitting information in digital form (in bytes) the number of bytes required to represent multimedia data can be very large. For example, to represent 2 minutes of uncompressed CD-quality music requires about 84 million bits. From this, one can gauge the time required to download all information from the Internet and Websites, the space required to store them, and the speed needed to handle all the data generated world over. Developing better transmission and storage techniques alone is not sufficient to meet the challenge. Morse code is an early example of data compression making use of sequences of (.) and (–) to represent letters that occur more frequently, e.g., e by (.) and a by (. -); and longer sequences to letters that occur less frequently, e.g. q by (—.-).

7.3.2. Expandability

If compression of information saves time for the user in grasping and communicating, the reverse process, i.e. expansion of a given concept stretches time for the user. Expansion is often used to impart literary and aesthetic values in presenting a text. Expandability enables compressed information easy to understand and appreciate by making it descriptive. Human mind employs distinctive information processes like description, illustration, explanation and comparison to impart expandability to information; and abstraction, extraction and coding to achieve compressibility of information. (Narayana, 2010)

7.4. COMMUNICATION

Study of the role of communication in the flow of information, in all its different forms, across time and geographical space from the time of its creation to the time it reaches the intended user is an aspect of information science. Information does not exist without communication, whereas knowledge does. Communication fulfills a social function of ensuring continuity in society through access to the experiences and ideas of the past, expressed in symbols for transmission across space and through time- a 'time-binding' function fulfilled through use of language, gestures and other social and cultural inheritances like rituals, ceremonies, and sacraments. Contrasting concepts of time have been the root of miscommunication in meetings between peoples from different cultures (Zahorian, 2008).

Society may be viewed as a closed system. Information, in the form of active and continuous communication taking place between the constituents in society, is the energy that maintains entropy. By increasing information and communication activities a society enhances its level of low entropy, which means high order in society (Neelameghan, 1979). Similarly societies depend upon cultural communication for maintaining homoeostasis, which increases its survival chances. Communication appears in different forms and takes place at different levels depending on the participants in the communication encounter and the nature of the domain.

Communication is affected by social, behavioural and economic factors. An example is the phenomenon of "publish or perish" operating in the field of scientific research. Publishing papers is taken as a critical standard to evaluate a scientist's research ability, prestige within the scientific community and outside, suitability to receive a research grant, and fitness for promotion in the career ladder.

Time and space elements, expressed or not, are inbuilt features of all formal communications, both oral as in a testimony and recorded as in a published paper. The time / space features appear as Domain time/Domain space and Disclosure time / Disclosure space. Domain time refers to the time/space when the event or action conveyed by the communication took place; and Disclosure time / Disclosure space refers to the time/date of communicating/reporting the event or action (Narayana, 2010). Disclosure time and Disclosure place are included in what is called metadocument in information science. History of civilizations is only interpreting and explaining events and actions occurring in domain times and domain space presented at a given disclosure time and disclosure place. A few of the time related issues in information science are briefly stated below.

7.5. TIME-RELATED PROPERTIES OF INFORMATION

Bibliometrics and informetrics are the sub-disciplines of statistics, They are applied for the development and use of a variety of measures to study and analyze several properties of information and documents, productivity of research, growth and obsolescence, citation studies and scattering (Ravichandra Rao, 2010). Scientiometrics developed as a field of study devoted to measuring and analyzing science is largely based on bibliometric methods. A citation based technique originally developed to assess the impact of theoretical physicist's relative quality, and now extended to new areas, is known as h-index or Hirsch number (Ravichandra Rao, 2010; p.61-66).

"Waiting time" is one of the standards used to determine the efficiency of an information system. There is always a waiting period in the working of a system from input to output stages. For a user of the system this may arise in any of the following stages: a) physical / communication access to the system; b) identification of and access to the system; c) use of the tools and techniques to search in the data base; d) selection and retrieval of the relevant information / surrogates; e) document / information delivery; and f) document/information usability (Neelameghan, 1979). The aim is to reduce the waiting period between the stages/agents in an information system – from the point of creation of knowledge to the final stage of access by the user. In the Internet world some of the above stated stages are taken care of by search engines.

Information Search Time. Individuals and organizations spend considerable time in searching for information in the course of their functioning. Search time may be defined as the time spent by individuals and organizations in searching for requisite information in an information source/ sources or systems. Search time includes the mental time invested on specifying the topic of search and its parameters, identification of information source/system to be tapped, actual time spent in going through the source and collecting the requisite information, and in the event of failure to mine the requisite information trying an alternative source. Search time cost is the money equivalent of the time invested on searching. This cost is independent of the cost of establishing and maintaining the information system / source, but includes the cost of failed searches.

Information search cost though not easy to calculate gets assigned to the overall information cost for the individual or the organization. There is also the problem of deciding how much information to be collected and knowing the correct information source. Inadequate information may lead to not meeting the goal, and too much information collection is wasting time and resources (Narayana, 2010). In some cases it may be necessary to decide how far back in time should the search be done. This may be related to depth of indexing the resources, availability of human resources to conduct in-depth search, etc., all of which are in turn related to investment of time and cost. The marginal returns from a long-drawn search may not be commensurate with the time (and resources invested for the purpose). Sometimes it works out more economical and prudent for an organization to meet its information needs through outsourcing than doing it in its own in-house information facility, thereby getting relatively better returns on its information search time.

Growth of Knowledge and Information. Knowledge is a growing phenomenon. Information is growing exponentially and is doubling every few years in most disciplines Exponential growth of information represents progress, but at the same time creates a number of problems in managing information and in optimizing use of information. Though Internet and Web site, and telecommunication technologies have provided efficient means of managing exponentially growing information - from creation to end use - have also created new serious problems. Internet is exploited by criminals, antisocial elements, and fraudsters. Dealing with copyright matters is quite tricky in the Internet communicated information. Studies have shown that exponential growth of literature must have ceilings; otherwise the growth would reach absurd conditions because of various constraining factors.

The most serious impact of information growth is on the time required to go through the exponentially growing stock of information in the world, to recognize and pick a small item of it that corresponds to one's requirement. In spite of the availability and development of numerous techniques - both human and machine, the complexities of collection, organization, and access is anything but smooth and perfect. For the not so thoroughly initiated, search for information is becoming something like searching for a pin in a haystack. The net result is that the searcher of information is forced to compromise with the integrity of information, which also means compromising with time. If new technologies have bridged time and space gap between the source of information and the seekers of it, the gulf between information that is relevant and effective and that of irrelevant and barren is increasing. In this situation, given the limitations of human mind, time takes its own toll in the form of information fatigue, cognitive syndrome, and knowledge phobia.

Dissemination and Diffusion of Information: Libraries, archives, museums, television, other news media play the role of disseminating information to the larger society. The environment of dissemination is general in nature. Whereas in diffusion of knowledge the target clients are specifically defined and the process is active. In diffusion time factor is very important. It is the element of time which distinguishes diffusion studies from other forms of communication. Researchers have developed several diffusion models to explain the process of diffusion and their effectiveness. The nature of diffusion differs from domain to domain and has social implications.

7.5.1. General Systems Approach

As an approach to understanding general systems theory, Boulding (1968) suggested the need "to look over the empirical universe and to pick out certain general phenomena which are found in many different disciplines, and to seek to build up general theoretical models relevant to these phenomena". Examples of such common phenomena are population dynamics, growth in time, relative growth, entropy, competition, information diffusion and so on.

Growth in Time: Exponential and Logistic Patterns : Patterns of growth in time showing isomorphism of law in different fields are the exponential and the logistic. Examples are the growth of science and technology, the increase in knowledge of the number of animal species, yeast cells in a controlled environment, the increase in publications on drosophila, in manufacturing companies, and so on. Isenson (1966) has shown that the rate of growth of knowledge over time can be expressed as an equation embodying various limiting factors such as the upper bounds imposed by natural law, political and social constraints, the contemporary level of scientific and technical understanding, and the ability of the scientific and technical communities to communicate their research findings. As several studies have shown, the amount of recorded knowledge in many scientific fields doubles, on the average, about every ten years, may be much faster now, but exponential growth must have ceilings because of various constraining factors, otherwise the growth would reach absurd conditions in the future.

The logistic growth curve may not be a normal S-shaped ogive. Historical time series analyses present variants of the curve as seen for example, in the growth of science and technology. Hunting fluctuations and the escalating curve are two examples. The constraining factors causing the variant shapes, and the implications to the sociology of science, have been the subject of study by De Solla Price (1963).

An example of the effect of some of these constraints e.g. time limit, in 'directed' research is the duplication of the discovery of antibiotics over a period of time and that it conforms to a modified Poisson distribution pattern (Neelameghan, Buche, Gupta, 1970). Forecasting of such duplication in a particular year, for instance, can be made using the square-cube law as well (Neelameghan, Buche, Gupta, 1970a)

Technology Growth: Boltzman Equation Model : Hartman (1966) found an analogy between nuclear chain reaction and technology growth. He has suggested that a model for the growth of technology can be derived directly from the Boltzman equation. Scientists and engineers are pictured as being immersed in a sea of information, the information being a necessary input for the generation of new information and ideas. His thermodynamic model simulates authors by stationary atoms and information carriers by electrons accelerated by an electrical field. A bit of information may collide with a researcher and trigger a new idea or it may not. As it is not possible to predict what piece of information will collide with which researcher and when and whether a particular collision will be 'elastic' or 'inelastic', the process is only statistically definable — providing an acceptable parallel for forecasting technology growth.

Information Diffusion: Heat Conduction Model : Avramescu (1975) describes a mathematical model for scientific information transfer on the analogy of heat-diffusion laws. The information-flow equation is similar to Fourier's. The information-diffusion process is a multiple chain, one scientific paper stimulating another. The elementary space elements are papers linked together by references and citations in a natural time sequence.

Idea Acceptance : Time-Dependent Models: Related to the concept of diffusion of ideas is the concept of acceptance of ideas by a target group. Morphet (1974) has presented a probabilistic model of time-dependent factors governing the acceptance of innovative ideas by industry. Within this framework is analyzed the timeliness factor that partially governs the receptivity to innovative ideas of industrial organizations. The model is based on the assumptions that: (a) even the best innovations do not diffuse by themselves; (b) the commitment of an organization to existing practice will favour innovation only when existing practices become obsolete. The occurrence of such obsolescence is identified with the concept of New Idea Point (NIP), the point in time at which the organization must have assembled a portfolio of innovative ideas in order to maintain continued existence. The reaction of firms to new ideas is associated with the organizational ambivalence - a tension between action and reaction, between need and conservatism - which must be resolved in order to meet the imperatives of the NIP.

Innovation Diffusion Models : Sociological Implications: Rogers and others (1968) have attempted a generalization of the mode of diffusion of research results in the 'market' over a period of time. An innovation is defined as an idea that is seen as new by an individual. Thus, innovations may include a genetically modified seed, a new weed spray (e.g., 2.4-D) among farmers, a drug among physicians, the intra-uterine contraceptive device among peasant women in India, an education technique among teachers, and so on. The crucial elements in the diffusion of innovations are deemed to be the new idea, the communication channels, the particular social system, and time. The element of time distinguishes the diffusion studies from other types of communication studies. Time is involved in the following ways: (a) the innovation decision period through which an individual moves from first knowledge of the innovation to persuasion of its usefulness, to its adoption, and to continued use; (b) the rate of adoption of the innovation in a social system, measured as the number of adopters per time period; and (c) the innovativeness or the degree to which an individual is relatively earlier than other members of his social system to adopt new ideas. Factors helpful in accelerating the diffusion and adoption of new ideas/innovations, such as attitudinal type variables, social relationships and strategies adopted by change agents, have also been studied (Elmore, 2001).

Hyperbolic equations : Fairthorne (1969; 2005) has reviewed the application of empirical hyperbolic distribution equations to bibliographic description and production. These include Bradford, Zipf, and Mandelbrot equations (see below). Goffman (1966) and Goffman and Newill (1964, 1967) have applied these empirical laws to the pattern of spread of epidemics or "epidemic model of the spread of scientific ideas as manifested by publication. Communication of ideas by recorded discourse is assumed to be governed by processes formally equivalent to communication of disease by microorganisms.... Information retrieval then corresponds to processes that enhance the spread of infection by favouring contacts of infectives with susceptibles.

Goffman-Newill use control system theory to study the growth, decay, and stability of such processes (over time), involves time explicitly, fitting the diffusion of innovations (and information about innovations) over a period. In recent years there have been many studies on models of the spread of epidemics which parallel or are applicable to the spread of ideas, innovations, messages, opinions, and information. For instance, Durett (2010) examines features of the spread of epidemics and information on a random graph. The models considered include random Boolean network, power-law degree distribution, epidemic model, and voter model. Newman (2002) looks at networks (contacts) in the spread of epidemics. Hollingsworth et al (2007) studied the role of travelers in the spread of respiratory diseases. Analogous to the bibliographic citation patterns and more frequently cited authors and periodicals in a particular domain, during a period of time that shows the relationship between authors and contributions of papers, a small proportion of travelers make 'disproportionately' more journeys than other travelers. They also tend to interact predominantly with other frequent travelers. This raises the potential to accelerate global spread. The pattern of and necessary conditions for, the spread of diseases have been studied not only among human beings but also in the plant and animal kingdoms. Goffman and Warren (1969) used their studies of mast cells and of schistosomiasis literature to investigate the Bradford distribution. They found that in both subjects the ratio of the periodicals to authors was almost the same (c.0.27), and the average output of authors over five-year periods did not vary over the entire period of eighty years covered by the bibliographies."

7.5.2. Acceleration of Innovation and Knowledge Obsolescence

The Second World War and the events that followed it provided a great impetus for scientific research, innovation and technological change, which in turn contributed in a large measure to socio-economic changes. Some of these innovations, particularly in information and communication technologies, made possible greater interaction among researchers and innovators, which in turn further accelerated the scope and pace of scientific and technological inquiry and innovation. It is estimated that half the total of scientific research in the United States has been done since 1965. Similarly, in the United Kingdom since the Second World War more than 50 per cent of all new products have been based on the more recent scientific and technological discoveries and innovations.

In addition to the expansion in the scope of scientific inquiry, there has been an appreciable acceleration in the rate at which new knowledge is put to use. This has also resulted in an acceleration of the rate of knowledge obsolescence. The United States Department of Commerce has estimated that prior to 1914 there was an average wait of 33 years between an invention and its application. By the Second World War, the time lag had dropped to 10 years and now it is even less. (*See also* Neelameghan, cited in Ranganathan, 1973).

Killingworth and Lynn (cited by Neelameghan, 1979, p.109) have found that: (a) the time lapse between innovation and its utilization in industry is much shorter for consumer products than for industrial products; (b) the time lapse is shorter for invention developed using government funds than for those using private funds; and (c) the technology diffusion and transfer time is reduced if the innovator himself / herself undertook the development. What are the implications? A major change that might have required five years to make a decade ago must now be completed in a shorter time period, if an organization, particularly an industrial or commercial organization, is to remain competitive. On the one hand management reaction time is constantly shrinking, while on the other hand each decision may involve more risk (because of a large number of variables, internal and external, to be taken into account, and is valid only for a shorter time span. Therefore, as reaction time diminishes, opportunity for gainful action may be lost, because preoccupied managers and policymakers may fail to reach out and grasp them.

7.5.3. Time in Collaborative Research

A feature of modern day big scientific research programmes (now extending to other fields too) is large-scale collaboration involving a number of laboratories and researchers, often spread across the globe, bound together by time.

7.5.3.1. Big Science, Massive Data

Today's big science is collaborative research generating mega-volumes of data over a period of years, which can only be handled by high-powered computers with large data storage capacity. For example, in the Large Hadron Collider (LHC) facility at CERN, Geneva, each experiment is a collaboration of over 5000 physicists from hundreds of research establishments spread all over the globe (Berman et al., 2003). This means that the activities involved in a given research programme is split into numerous segments and sub-segments. The segmented tasks are assigned to various participants to be completed according to the set time-plan. Inherent in the plan is time-distribution as well as time-coordination. The magnitude of time and information coordination involved in large collaborative research can be seen from the example of research conducted at CERN. Experiments at CERN generate 40 terabytes of data every second - orders of magnitude more than can be stored or analyzed. The result is that scientists collect what they can and let the rest dissipate into the ether (Data, data, 2010). One can imagine the variety and the volume of data that will be produced by each such experiment. Other fields of research which generate extremely huge amounts of data include astronomy, biological/genetic research, medical/health monitoring, climate/ geological studies, and mechanical equipment function monitoring. Study of the data can result in important new insights about life and nature. We are living in the era of "big data". Sensors and other devices are now available to digitize huge amounts of data, analyze them, aggregate them, and put to use in a relatively short time (Data, data, 2010).

Mega-data generating research has changed the traditional pattern of scientific communication, which is built around authors (mostly individuals and small groups), who create data; publishers of scientific journals; libraries playing the role of *curators*; and scientists using the published data. The pattern of communication in the mega-data world needs a new model. In the new model, "the author, publisher, and curator roles are clearly present in data-intensive science, but they are performed in different ways" (Szalay and Gray, 2004, p.98). In the new structure, the author is identified with collaborations; the data is published through Web-based archives; and during the life time of a project the curatorial responsibility lies with the projects themselves. Later on, data considered valuable are transferred to national data archives. Project data management involves custom coordination of distributed data, their consolidation, standardization of nomenclature; design of data forms and web site tuning. Raw instrument data need to be transformed into calibrated and catalogued data, and even customizing. Data-driven mega scientific research has to handle scientific data in the range of terabytes (2 to the power 40 bytes) to exabytes (2 to the power 60 bytes) - to create a five letter English word takes 40 bytes.

7.5.3.2. Time in Business Data Growth

A notable feature of the 21st century global economy is the astronomical quantity of business data generated worldwide, and its conversion into useful business information. As Craig Munde of Microsoft Corporation observes that what we are seeing in this growth is "the ability to have economies from around the data - and to me is the big change at a societal and even macroeconomic level..." (Data,

data, 2010, p.4). A large part of this business information is accounted for by data gathered by firms relating to spending habits, information inclinations, and behavioural patterns of their customers or subscribers by probing into the customers' online transactions record, often surreptitiously. The collection of such information forms a gold mine of "business intelligence" of immense value to the firm, and a commodity to many others. The search engine Google automatically recycles data extracted from user interactions, derived as a byproduct during a period, to improve its service or create new products (Data, data, 2010).

The amount of information to be handled, the time-coordination, consolidation and updating involved in this process can be gauged from the information processing system evolved by the marketing giant Wal-Mart. Wal-Mart having world-wide interest, handles more than 1 million customer transactions every hour, feeds databases estimated at more than 2.5 petabytes - the equivalent of 167 times the books in the US Library of Congress. Another example of compression of time is that of Hadoop, a credit-card company, which processes 73 billion transactions in just 13 minutes which would have taken earlier a month following a traditional method. (Data, data, 2010). Digital information in India is projected to grow from the present 40,000 petabytes to 2.3 million petabytes by 2020 - a growth of 60 times in just 10 years (Chugh, 2010).

It is evident that the amount of data generated today cannot be processed manually as there is neither the time nor the capability to handle and manipulate this data. Only machines can handle the task. Roger Bohn of the UCCD explains "information created by machine and used by other machines will probably grow faster than anything else.... This is primarily 'database to database' information - people are only tangentially involved in most of it" (Data, data, 2010, p.5). Managing mega data has lead to the emergence of a new kind of professional, "the data scientist, who combines the skills of software programmer, statistician and storyteller/artist to extract the nuggets of gold hidden under mountains of data" (Data, data, p.4). This shows the importance attached now to the post of Chief Information Officer

(CIO) in business and industrial enterprises. As The Economist report concludes "in the age of big data, computers will be monitoring more things, making more decisions and even automatically improving their own processes - and man will be left with the same challenges he has always faced. As T.S. Eliot asked "Where is the wisdom we have lost in knowledge? Where is the knowledge we have lost in information?" (Data, data, 2010, p.4). This is an enormous leap taken by man to "de-past" past time by real-time collection and processing of data and its conversion to information with the help of advanced technological processes developed by drawing from computer science, information technology, artificial intelligence, linguistics, mathematics and a few others.

7.5.4. Statistical Patterns in Information Phenomena

Since the 1960s, in bibliometrics, two classes of empirical laws have been used fairly frequently. These are the Zipf and Bradford laws (Zipf, 1949; Bradford, 1948) both of which are hyperbolic distributions. One or the other of the laws is used, depending upon whether one's interest is in vocabulary, periodical literature, physical access time, the rate of diminishing returns in bibliographical search, or the cumulative yield from a given input. The time factor involved here is in the explanation of the observed relation in many fields on the basis of the general principle of least effort.

7.5.4.1. Document Life Parameter

It has been observed that if all the citations in a single issue of a periodical, or a volume of it for a particular year, are sorted according to the date of the cited documents, then the number falls off rapidly as we move back into time. Since 1960 several studies of this type in different subject fields have been done (Burton and Kebier, 1960). From these studies several time-related parameters of a periodical's life have been derived, such as 'half-life'- the time period, actual or expected, during which half the total use of an individual item constituting a literature has been, or is expected to be, made.

The values of such parameters can be used for guiding library management decisions, such as: maintenance of library collections (periodicals, books, etc.), withdrawal of documents from active circulation, planning of acquisition programmes within a hierarchy of library systems, and identifying significant research in a field.

7.5.4.2. Impact Factor and Indices of Performance

In recent years there have been an increasing number of studies relating to the measurement of productivity of authors, institutions, journals, etc., and various methods of computing the indices for these factors. There is yet no universally accepted method and index of productivity / performance applicable to all disciplines. Nevertheless, in all these measures, whether relating to individuals or institutions, journals or web resources there is a time parameter involved in the computation. The study may be diachronous (forward in time) or synchronous (backward in time).

7.5.5. Research Communication: Sociological and Behavioural Aspects

7.5.5.1. Overemphasis on Speed?

Timely availability of relevant and reliable information and data is important to the researcher, the production engineer, the manager and planner, as an aid to solving problems, making gainful decisions, minimizing the chances of unnecessary duplication of effort, and triggering new ideas. With the growing demand for information and data, there has been a phenomenal growth in primary sources such as books, periodicals, reports, conference proceedings, and so on. In finding ways and means of moving the information from the point of generation to points where it will be of use, the need for speed of communication has been emphasized. For instance, reducing the time lag between submission of a paper to a periodical and its publication has been an important consideration. Growth in the number of primary periodicals, publication of short communications, minimizing the editorial work, and the adoption of fast printing techniques are some of the results of such considerations. There have been salutary effects as well as the undesirable effects of the ever-increasing quantity of information that swamps the user at an apparently ever-increasing speed. Application of emerging information and communication technologies (ICT) to the different processes and steps in journal publications has helped in shortening (time) the author submission of paper, refereeing and editing, and publishing and distribution. These days authors / researches even tend to place their write-up directly on a website, or present the paper to a conference audience via the YouTube. There is a mushrooming of journals inviting authors to publish in 'their' periodical promising very little time-lag between submission of the paper and its appearance in the periodical!

In the pure sciences the supposed wastage of resources through duplication of effort is somewhat exaggerated. Simultaneous discovery arising from concurrent work on the same problem by different researchers is not a bad thing in itself. It provides for a good means of confirmation of the findings or discoveries; the routes taken to the discoveries by the different researchers may be different. Experiments may be repeated to improve techniques or increase accuracy of data. Also, the 'speed of research' varies at different stages, from discipline to discipline, and the 'speed of supply of information' should correspond to this variation in demand (Ziman, 1969).

The rate of proliferation of primary publications and of information exchange media has gone on unabated. The 'publish or perish' syndrome is bound up with many questions in the sociology of science (i.e. the achievements of a scientist, especially the newcomer to a field, being estimated on the basis of the number of papers published by him / her, the economics of research and the rat-race for grants, the behavioural aspects of researchers to 'scoop the field' as in competitive business to ensure priority of 'discovery', which in turn could bring more projects and grants, the economics of the information industry, and so on.

The sociological and human aspect of science also throws some light on the motivations of the scientist for quick announcement and speedy communication of his/her work. The motivation is to establish and maintain intellectual property rights. Duplication of discoveries is a widespread phenomenon especially in directed research environments (cf. Neelameghan, Buche, Gupta, 1970).

The issues discussed above relate to communications in 'pure' sciences field. The situation is somewhat different in the applied and development research areas. Here one seeks the best available solution to a particular problem, and the rewards of success or cost of failure may be measured in millions of rupees. Timely information and data helpful in making gainful decisions are vital in competitive industrial research. Here too, however, when users are research managers, production managers, etc., the information system should not only be capable of picking out quickly the most relevant data and information, but also present them in a manner - for example as digest, technical note, trend analysis with charts and graphs - that will immediately show the relevance of the information to the activities of the organization or the individual concerned. Such repackaging of the information and data would facilitate the reaction of the individual to the perceived situation in the context of a rapidly decreasing management reaction time to environmental changes.

It may be noted that "hindrances to the immediate spread of new knowledge (in applied and development research) are not then so much in the machinery of publications, but rather barriers of secrecy deliberately raised around industrial and military research. The communication system of technology is quite different from that of pure science, having different ends, different norms, and altogether different standards of morality." (Ziman, 1969).

7.5.5.2. Behavioural Aspects

The amount of information processed by businessmen, consumers, stockholders and other 'economic actors' influences their behaviour on the stage of the economic system. Using such considerations and looking upon information and knowledge (both embodied and easily accessible and tacit knowledge) as key resources, are seen to influence the pattern of thought in several domains - economics, politics, policy making, governance, and so on (Arrow, 1996). Economists have been developing an economics of information, and over the past decades it has ranged over a wide spectrum of activities in the information field - from the study of specific activities of information systems to a reconsideration of the role of information in the economic system and the welfare economics of symbol manipulation.

8

Time in Historical and Some Evidence-Based Studies

Natural evolution is a slow process, changes taking place over millions of years. Societal evolution and changes take place in relatively shorter period of time. The pace of changes in society is further pushed up with the advent of information communication technologies (ICT). The impact is felt more frequently than ever before.

In an earlier chapter we briefly discussed the social function of information communication - survival in time; ensuring continuity in society; time-binding achieved through symbolic communication; greater societal cohesion and conflicts possibilities; and societal homoeostasis and entropy.

We have already mentioned that 'Time' is an ubiquitous element associated in a variety of ways in human activities and, therefore, the related records describing them. We have also seen that the different philosophies and cultures conceive of, identify and/ or measure time in different ways.

In almost all studies more particularly in the sciences proof / evidence is very basic. There are several subjects, such as History, in which time - past, present, and even envisioned and prophesied future

– is studied, to determine when an event occurred (e.g. World War I, Mahatma Gandhi's Dandi March, etc.) or may occur; what was the duration of the War of Roses and similar questions are raised and answers found – a primary function of information systems and services. In the following sections we briefly consider a few subjects in which 'Time' of certain occurrence and its recording and inference drawn on the basis of records of time.

History is defined variously: (1) the aggregate of past events; the continuum of events occurring in succession leading from the past to the present and even into the future; all of human history; (2) records such as documents of different genres, a phonograph record or a photograph, providing permanent evidence of or information about past events in the form of written record - memorabilia, file, data file, account, includes working papers that is records kept of activities involved in carrying out a project; (3) the discipline that records and interprets past events involving human beings (WordNet 2.1). Viewed from these perspectives every subject, macro or micro, has a historical aspect.

Historiography is discoursing the writing of history, usually based on evidences – archaeological and documentary in any medium, and oral (passed down the years). It also connotes a collection of historical documents.

Archaeology "is the study of human activity in the past, primarily through the recovery and analysis of the material culture and environmental data that humans have left behind, which includes artifacts, architecture, biofacts and cultural landscapes (the archaeological record). Archaeology uses a wide range of different procedures; it has a scientific component and a humanities component. In some countries, such as, USA, it is considered as a branch of anthropology, but in Europe it is considered as a separate discipline (Wikipedia).

Archaeology studies *human history* from the development of the first stone tools 3.4 million years ago in eastern Africa up until recent times. Archaeological studies are important "for learning about prehistoric societies, when there were no written records for historians to study, making up over 99% of total human history, from the Palaeolithic until the advent of literacy in any given society. The goals of archaeology cover a wide spectrum ranging "from studying human evolution to cultural evolution and understanding culture history." Archaeology also sheds light on many of humanity's technological advances at different periods of its development, e.g. the making and use of fire, the development of stone tools, the discovery of metallurgy, the beginnings of religion and agriculture. Without archaeology, it would be practically impossible to know about the use of material culture by humanity before writing was invented. Concept terms with paleo- or archeo- prefix fall in this category. The early periods also get such names as stone age, iron age, prehistoric time, ice age, etc.

The methods of study used include survey, excavation and analysis of data collected to learn more about the past. It is a cross-disciplinary study drawing from anthropology, history, art-history, classics, ethnology, geography, linguistics, semiology, geology, physics, information science, chemistry, statistics, paleontology, paleo-ecology, paleozoology, paleoethnobotany, and paleobotany. Its sub-disciplines include maritime or underwater archaeology, feminist archaeology and archaeo-astronomy, and numerous different scientific techniques have been developed to aid archaeological investigation.

In more recent times, archaeologists use such methods as remote sensing, aerial survey, geophysical survey, field survey, excavation, using different tools and devices, and applies statistical methods in data analysis. Increasingly computer graphics and modeling (virtual archaeology) are being used. Historical archaeology, ethno-archaeology, experimental archaeology, archaeometry, cultural resources management are some of the academic sub-disciplines of archaeological studies.

Fossils: Closely related to archaeological studies is study of fossils. Fossils "are the preserved remnants or traces of animals (also known as zoolites), plants, and other organisms from the remote past times. A *fossil record*' is the totality of fossils, both discovered and undiscovered, and their placement in (fossil-containing) rock formations (fossiliferous) and sedimentary layers (strata).

The study of fossils across geological time, how they were formed, and the evolutionary relationships between taxa (phylogeny) are some of the important functions of paleontology. Such a preserved specimen is called a "fossil" if it is older than some minimum age, usually the arbitrary date of 10,000 years ago. Hence, fossils range in age from the youngest at the start of the Holocene Epoch to the oldest from the Archaean Eon, up to 3.4 billion years old. The observations that certain fossils were associated with certain rock strata led early geologists to recognize a geological timescale in the 19th century. The development of radiometric dating techniques in the early 20th century enabled geologists to determine the numerical or *"absolute" age* of the various strata and thereby the embedded fossils.

Like extant organisms, fossils vary in size from microscopic, such as single bacterial cells only one micrometer in diameter, to gigantic, such as dinosaurs and trees many metrer long and weighing many tons. A fossil normally preserves only a portion of the deceased organism, usually that portion that was partially mineralized during life, such as the bones and teeth of vertebrates, or the chitinous or calcareous exoskeletons of invertebrates. Preservation of soft tissues is rare in the fossil record. Fossils may also consist of the marks left behind by the organism while it was alive, such as the footprint or feces (coprolites) of a reptile. These types of fossil are called trace fossils or *ichnofossils*, as differentiated from *body fossils*. The past life may leave some markers that cannot be seen but can be detected in the form of biochemical signals; these are known as *chemofossils* or biomarkers. Estimating the dates of fossils is geochronology.

Paleontology's goal is to map out how living things have changed over time. A great hurdle to attaining this goal is the difficulty of working out how old fossils are. Beds that preserve fossils typically lack the radioactive elements needed for radiometric dating. This technique is the only means of giving rocks greater than about 50 million years old an absolute age, and can be accurate to within 0.5% or better. Radiometric dating is based on the rates at which various radioactive elements decay are known, and so the ratio of the radioactive element to the element into which it decays shows how long ago the radioactive element was incorporated into the rock. Radioactive elements are common only in rocks with a volcanic origin, and so the only fossil-bearing rocks that can be dated radiometrically are some volcanic ash layers.

Consequently, palaeontologists usually rely on stratigraphy to date fossils. Stratigraphy is the science of deciphering the "layer-cake" that is the sedimentary record. Rocks normally form relatively horizontal layers, with each layer younger than the one underneath it. If a fossil is found between two layers whose ages are known, the fossil's age must lie between the two known ages. Because rock sequences are not continuous, but may be broken up by faults or periods of erosion, it is very difficult to match up rock beds that are not directly next to one another. However, fossils of species that survived for a relatively short time can be used to link up isolated rocks: this technique is called *biostratigraphy*. Such index fossils must be distinctive, be globally distributed and have a short time range to be useful. However, misleading results are produced if the index fossils turn out to have longer fossil ranges than first thought. Stratigraphy and biostratigraphy can in general provide only relative dating which is often sufficient for studying evolution. However, this is difficult for some time periods, because of the problems involved in matching up rocks of the same age across different continents.

Family-tree relationships may also help to narrow down the date when lineages first appeared.

It is also possible to estimate how long ago two living clades diverged– i.e. approximately how long ago their last common ancestor must have lived – by assuming that DNA mutations accumulate at a constant rate. These "molecular clocks", however, are fallible, and provide only a very approximate timing: for example, they are not sufficiently precise and reliable for estimating when the groups that feature in the Cambrian explosion first evolved, and estimates produced by different techniques may vary by a factor of two

(Wikipedia).

There are limitations to fossil records and their use.

Genealogy is the study of families and the tracing of their lineages and history. Such studies use archival records, historical records, oral traditions / oral history, genetic analysis, and other sources to get data and information about a family. From the data collected such information as relationship / kinship and pedigrees among its members may be derived. Tree charts and written narratives are used to display the relationships among the members.

Family history studies help to identify one's position and place in the hierarchy; another motivation may be to identify one's place and of the family in the larger historical picture; and it may give a sense of responsibility to preserve the past for future generations. Claims to property may be another reason

A variety of records are used in genealogical research. The researcher needs to examine and understand when and how the records were created, what information is included in them, and how and where to access them. The following are some of the types of records used in genealogical research (Wikipedia):

Vital records

- Birth records
- Death records
- Marriage and divorce records

Adoption records

Biographies and biographical profiles (e.g. *Who's Who*)

Census records

Church records

- Baptism or christening
- Confirmation
- Bar or bat mitzvah
- Marriage
- Funeral or death

- Membership

City directories and telephone directories

Coroner's reports

Court records

- Criminal records
- Civil records

Diaries, personal letters and family Bibles

Emigration, immigration and naturalization records

Hereditary & lineage organization records

Land and property records, deeds

Medical records

Military and conscription records

Newspaper articles

Obituary notices

Occupational records

Oral histories

Passports

Photographs

Poorhouse, workhouse, almshouse, and asylum records

School and alumni association records

Ship passenger lists

Social Security or its equivalent and pension records

Tax records

Tombstones, cemetery records, and funeral home records

Voter registration records

Wills and probate records

Regarding dates recorded in any of the records mentioned above one must exercise caution because dates are more difficult to recall years after an event, and are more easily mistranscribed. Therefore, one should determine whether the date was recorded at the time of the event or at a later date. Dates of birth in vital records or civil registrations etc are likely to be more accurate than other genealogical dates, as in many countries they are recorded at the time the event occurred. There are several reasons for a person to give / record a different date than the actual date. .There has been changes in the popular calendars also.

Europe : In England and Germany, for example, such record keeping began with parish registers in the 16th century. As details about more members of the population were recorded, it was possible to follow a family line. Major life events, such as births, marriages, and deaths, were often documented with a license, permit, or report. Genealogists locate these records in local, regional or national offices or archives and extract information about family relationships and recreate timelines of persons' lives. In Ireland, genealogical records were recorded by professional families of *senchaidh* (historians) as late as the mid-17th century, when Gaelic civilization died out. A good example of such record is Leabhar na nGenealach. The Great Book of Irish Genealogies, by Dubhaltach MacFhirbhisigh (d. 1671), published in 2004.

In China, India and other Asian countries, genealogy books are used to record the names, occupations, and other information about family members, with some books dating back hundreds or even thousands of years. In the eastern Indian state of Bihar, there is a written tradition of genealogical records among Maithil Brahmins and Karna Kayasthas called "Panjis", dating to the 12th century CE. Even today these records are consulted prior to marriages (Wikipedia).

Astronomy, one of the natural sciences, deals with the physics, chemistry and evolution of celestial objects, such as moons, planets, stars, nebulae, and galaxies; the physics, chemistry, and evolution of such objects. Also included are studies on phenomena that originate outside the atmosphere of Earth - supernovae, gamma ray bursts, and cosmic background radiation. Cosmology, a related subject, covers studies on the universe as a whole.

Astronomy is one of the oldest sciences. Prehistoric cultures have left behind astronomical artifacts such as the Egyptian monuments and Nubian monuments. Early civilizations such as the Babylonians, Greeks, Chinese, Indians, Iranians and Maya had carried out methodical observations of the night sky. With the invention of the telescope astronomy began developing into a modern science. Historically, astronomy has included such disciplines as astrometry, celestial navigation, observational astronomy, and the making of calendars. In modern times professional astronomy is often considered as synonymous with astrophysics.

During the 20th century, the field of professional astronomy split into two branches: (1) observational and (2) theoretical. Observational astronomy is focused on acquiring data from observations of celestial objects, which is then analyzed using basic principles of physics. Theoretical astronomy is oriented towards the development of computer or analytical models to describe astronomical objects and phenomena. The two fields complement each other, with theoretical astronomy seeking to explain the observational results, and observations being used to confirm theoretical results. Specific subfields include: Solar astronomy, Planetary science, Stellar astronomy, Galactic astronomy, Extragalactic astronomy, and Cosmology. Observational astronomy includes: Radio astronomy, Infrared astronomy, Optical astronomy, Ultraviolet astronomy, X-ray astronomy, Gamma-ray astronomy, Fields not based on the electromagnetic spectrum, and Astrometry and celestial mechanics.

An archive is a collection of historical records, or a depository containing historical records and documents. Archives contain primary source documents that have accumulated over the course of an individual's or organization's lifetime, and are kept to get information about an organization.

In general, archives consist of records that have been selected for permanent or long-term preservation on grounds of their enduring cultural, historical, or evidentiary value. Archival records are normally unpublished and almost always unique, unlike books or periodicals for which many identical copies exist. Archival collections may be found within library buildings, but archives have distinct functions and organization, from those of libraries.

The practice of keeping official documents is very old. Archeologists have discovered archives of hundreds (and sometime thousands) of clay tablets going back to the 3rd and 2nd millennia BC in sites like Ebla, Mari, Amarna, Hattusas, Ugarit, Pylos. These discoveries have been fundamental to know ancient alphabets, languages, literatures and politics.

Archives were well developed by the ancient Chinese, the ancient Greeks, and ancient Romans (who called them *Tabularia*). However, they have been lost, since documents were written on organic materials like papyrus and paper. However, many archives established since Middle Age by churches, kingdoms and cities survive and often have kept their official status uninterruptedly even in the present time. They are the basic tool for historical research.

The French National Archives, which holds perhaps the largest archival collection in the world, with records going as far back as 625 BCE, were created in 1790 during the French Revolution from various government, religious, and private archives seized by the revolutionaries. Modern archival philosophy and principles and practices have roots in the French Revolution (Wikipedia).

Apart from 'historical' studies briefly described above, subjects such as Medicine (healthcare) and Law depend on records of time. For instance, in the inpatient's record among other data the date of birth, age, the date and time of admission, duration of treatment, prescribed drug dosage and timings, the date and time of surgery, duration of the surgical intervention, the date and time of discharge, the date and time of re-admissions, etc. and recording of time-related data, such as, heart-beat, and pulse rate are required for subsequent use / consultation, clinical discussions, etc. as necessary, including when forensic (medico-legal) questions arise. Such case records data are also used in research, data analysis, statistical studies, training and forensic investigations (criminology).

The value of such records is the basis of evidence-based decision making in healthcare.

The Cochrane Collaboration, founded, in 1993, is an international not-for-profit and independent organization, dedicated to making up-to-date, accurate information about the effects of health care readily accessible worldwide. It produces and disseminates systematic reviews of health care interventions and promotes the search for evidence in the form of clinical trials and other studies of interventions. This collaboration is named after the British epidemiologist, Archie Cochrane. (http://www.cochrane.org/about-us/evidence-based-health-care) *Bulletin of the NYU Hospital for Joint Disease* (2009), 67(2); 198-205

The records of findings of population-based research form the foundation of evidence-based medicine. It aims to use the experience of a population of patients reported in the research literature to guide decision making in practice. This practice of evidence-based medicine (EBM), which later evolved to evidence-based health care (EBHC), requires the application of population-based data to the care of an individual patient. In the past, many have relied on the experience of physicians or other health care workers to make decisions about therapy. In the current information era, this approach is inadequate as health care workers find themselves unable to cope with the influx of a huge variety of new information, from the irrelevant to the very important. Therefore, Evidence-based decision making gradually emerged as a solution to integrate the best research evidence with clinical expertise and patient values and expectations as practiced by the individual health care provider. The EBM/EBHC-related principles and methodologies, concepts of hierarchy of evidence, meta-analyses, confidence intervals, study design, and so on are becoming more widely accepted and practised. Cochrane Library usage data: "*Every day someone, somewhere searches The Cochrane Library* ***every second****, reads an abstract* ***every two seconds*** *and downloads a full-text article* ***every three seconds***".

Law : The practice of law is largely evidence-based and case reports, which contain the year, date and time of occurrence of events / happenings are researched / investigated and recorded (Sri Ram, 2007).

9

Ontology and Representation of Time in Knowledge Organization Tools Used in Information Systems

9.1. ONTOLOGICAL APPROACH TO REPRESENTATION OF TIME

Webster's dictionary defines ontology as:

a) "a science or study of being: specifically, a branch of metaphysics relating to the nature and relations of being; a particular system according to which problems of the nature of being are investigated; first philosophy; and

b) "a theory concerning the kinds of entities and specifically the kinds of abstract entities that are to be admitted to a language system" (Merriam-Webster online).

According to Mills (2000) ontology is the philosophical position that underpins our theoretical assumptions and ideas about existence. Ontological assumptions frame not only our understanding of reality, but direct our epistemological approach. An epistemological standpoint is how we come to gain knowledge.

From the above definitions it is seen that ontology has philosophical and epistemological dimensions. While the philosophical position addresses social and cultural knowledge from the standpoint of temporality, the epistemological standpoint describes representation and organizational aspects of cultural and social concepts for later retrieval, which falls under the domain of information science and information systems.

In designing systems for representing, organizing, and retrieving information it is ontology taken in its epistemological sense that reveals the first realities which deal with the fundamental categories of object, state of affairs, part, whole, and so forth, as well as the relations between parts and the whole and their laws of dependence (Albartazzi, 1996). Here ontology is "an explicit specification of a conceptualization" (Gruber, 1993). It is how the term is used in artificial intelligence and knowledge representation.

In computer science and information science, ontology is a formal representation of a set of concepts within a domain and the relationships between those concepts. It is used to reason about the properties of that domain, and may be used to define the domain. In theory, ontology is a "formal, explicit specification of a shared conceptualization", that is, an ontology provides a shared vocabulary, which can be used to model a domain — that is, the type of objects and/or concepts that exist, and their properties and relations. Ontologies "are used in artificial intelligence, the Semantic Web, systems engineering, software engineering, biomedical informatics, library science, enterprise bookmarking, and information architecture as a form of knowledge representation about the world or some part of it."(Wikipedia). Ontology in Artificial Intelligence (AI) refers to an engineering artifact, constituted by a specific vocabulary used to describe a certain reality, plus a set of explicit assumptions regarding the intended meaning of the vocabulary words. This set of assumptions has usually the form of a first-order logical theory, where vocabulary words appear as unary or binary predicate names, respectively called concepts and relations. In the simplest case, ontology describes a hierarchy of concepts related by subsumption relationships; in more sophisticated cases, suitable axioms are added in order to express other relationships between concepts and to constrain their intended interpretation (Corazzon, 2010). The creation of domain ontologies is also fundamental to the definition and use of an enterprise architecture framework. (WordNet 2.1). From another point of view ontology is considered as an enabling technology for the Semantic Web. It assists in stating the connotation / meaning of the data that one may generate or use or share with other information seekers. Here ontology enables people to state what they mean by the terms used in data / information that they generate, share, or use. Folksonomies are an emergent phenomenon of the Social Web. They arise from data about how people associate terms with content that they generate, share, or use (Gruber, 2008). Gruber further adds that as the Semantic Web matures and the Social Web grows, there is greater value in applying Semantic Web technologies to the data of the Social Web.

To overcome confusions that may be caused by socio-cultural differences in expressing time there is a need for ontology of Socio-cultural Time Expressions (SCTE). The purpose of such an ontology is "to explicitly and formally describe these context dependent temporal expressions. This is done by defining appropriate hierarchical categories to which the expressions will be assigned. The ultimate goal is to be able to develop context aware temporal applications for the Web such as calendars or appointment schedulers" (Wennerberg and Schulz (2010, p.64).

Ontology of SCTE for Web Applications: A predominant characteristic of the information society is that a large number of individuals and organizations conduct their business and interactions through the Internet and the World Wide Web or Web. Hence, the users of the Web with their diversified socio-cultural background are also subject to the same or similar problems caused by the differing socio-cultural time in their interactions, particularly involving temporal information. This has necessitated the need for developing software tools for the Web which is capable of expressing different socio-cultural time concepts. SCTE are a subset of natural language expressions that are related to nations, religions, business life and education (Wennerberg and Schulz, 2010).

A number of independent time ontology for the Web exist (Allen et. al, 2002; Beule, 2003). Time-DAML is one of the well known ontology of time that is designed to provide annotations for the contents of Web resources. There are also a number of broad-based culture ontologies. Most of them provide models for cultural heritage systems (Doerr et al., 1999), for cultures in different enterprises and organizations (Abou-Zeid, 2003) or for culture specific language terms (Agnesund, 1997). However, these ontologies treat time and culture as concepts that are independent of each other, whereas Web applications need interdisciplinary ontologies, that is socio-cultural temporal ontologies that address the concepts together. In this context, Wennerberg and Schulz argue that "socio-cultural temporal information needs to be described formally and explicitly so that it can be processed by dependent software applications. This can help avoid potential conflicts between temporal software applications, e.g. Web calendars that were designed to meet different cultural needs and expectations" (Wennerberg and Schulz, 2010, p.63). Towards meeting this goal, they have proposed an ontology-based model to formally, explicitly and unambiguously describe socio-cultural time expressions, which they say is appropriate not only because its level of formality (hence machine-process ability) but also it gives Web access to the knowledge contained in it. To enable Web access, their ontology of socio-cultural time expressions is implemented on an OWL model using the methodology MENTHONTOLOGY for engineering process. The ontology follows hierarchical structure. Categorization of time items at the first level is based on: (1) Time Expressions denoting Time Periods, with the sub-category of Time Periods that can be anchored on the time-line/calendar, (2) Socio-cultural Time Periods, (3) Socio-cultural Time Periods Related to Religions, (4) Socio-cultural Time Periods having a duration longer than One Day, and (5) Socio-cultural Time Periods with Duration of One Day. Each category is further sub-categorized. A concept dictionary is created that lists all the concepts, properties and relationships that are to be included. The scope of the ontology is restricted to a subset of socio-cultural time that is related to nations, religions and business life, and education as seen in Turkish, US and Israeli nations, and the Islamic, Christian and Jewish religions. An important issue in the construction of socio-cultural ontologies on which there is no general consensus so far is about modeling of natural language expressions, particularly lexical and non-lexical types of information.

9.2. INFORMATION SYSTEMS

An information system, whether it is automated or manually operated is a time-delay mechanism. Its archive, repository, database, as also the Web, library, etc. are collections or hold collections of embodied data / information / knowledge in documents or other media created and expressed by 'authors', intended for use in the present and in the future. These information resources are processed, organized, catalogued and indexed and manual or machine-assisted retrieval-enabled so as to satisfy the Laws of Library Science - every item of information is used and every user gets his/her piece of information through pinpointed, expeditious and exhaustive service, saving the time of the user and of the information resource managers. Such collections grow and their contents change over time. Thus most information systems are dynamic. The arrival of Internet and Web sites has revolutionized the entire field of information creation, processing and delivery. Whatever be the type of information system, the fundamental task in designing the system is how we represent knowledge/ information. In other words it means how we represent concepts (Neelameghan, 1979).

Information systems and services provide accesses to information and/or deliver it. The primary task involved here is how information which is made up of interlinked concepts are expressed and represented in the system and communicated. The process of representation of concepts and their relationships includes standardization of terms, showing their interrelationships, and indicating the scope of each expression or group of terms. This practice is well established in designing of formal information systems. However, a rapidly expanding dimension of information servicing and use seen across the globe is the exchange of information-between one-to-one, and between one-to-many carried out through e-mail and social networks like facebook, blog and twitter. Here, the process of expressing ideas takes place in a purely informal way as the exchange of information takes place directly between the parties to engagement in cyberspace. In this the context is very important; and the context is often socio-cultural. Hence one of the central issue of study in information system design is how concepts and ideas relating to time is represented in information systems and in informal exchanges, given that time is linked with the socio-cultural environment enveloping the time or period in question.

At any given period of time, the cultural and social life of a society is greatly influenced by its philosophy, religion, and belief systems. It is in this milieu concepts and knowledge develop. Thus, philosophical and religious realities of the period provide the essential background for studying the cultural and social knowledge in a given society as affected by time. This is particularly so in the case of East Asian countries such as India, whose cultures and philosophies are rich with concepts relating to Time because of their long cultural heritage. (Evers and Gerke, 2003).

9.3. KNOWLEDGE ORGANIZATION TOOLS (KOT)

Knowledge Organization Tools (KOT) include schemes for subject classification, taxonomy, ontology, thesaurus, subject (heading) list, self-organizing system, etc. Concepts falling in the domains for which a KOT system is designed are represented by words of a natural language (English, French, Hindi, Tamil, etc.) and or by codes e.g. in classification systems using international numerals, numerals of a particular language, alphabets of a selected language (e.g. Hindi, Kannada), and punctuation marks used as indicator digits. A particular system may use a combination of words of a natural language and/or codes and/or punctuation marks. KOTs are applied mainly to organize embodied knowledge (documentary and non-documentary forms) about a subject or subjects. They are also used in information / document retrieval, e.g. for vocabulary management. A subject domain comprises of a number of concepts related to each other – hierarchical relationship and non-hierarchical associative (or lateral) relationship. Sub-division of a subject and overlapping subjects contain common concepts related to each other in similar patterns / types of relationship.

An initial step in building a KOT is to create the corpus of terms adequately representative of the concepts in the domain concerned. Identifying and selecting the sources, identifying the nature of relationships among the terms, organizing the terms so as to represent the relationships among the terms to create schedules (classification scheme, thesaurus, taxonomy), creating hyperlinks among the terms, etc. If the domain is specific (and narrow) individualizing elements so as to contextualize terms may not be necessary. If the domain is broad to comprehend multiple specific domains, then such individualization may become necessary. This may be achieved by enumerating the broader and narrower terms in their respective proper sequence.

A single KOT does not adequately indicate the multiple and complex relationships of the multi-dimensional universe of knowledge; an integrated or combined use of two or more KOTs can provide a better picture. The problems are compounded when multilingual information resources are to be organized. (Jorna and Davis, 2001; Neelameghan and Raghavan, 2007).

9.3.1. Representation of Semantic Relationship in a KOT

In non-faceted enumerative classification schemes essentially hierarchical relationships among concepts are represented. In a faceted scheme, hierarchical relationships among concepts in each facet may also be represented. In the synthesized class number (or terms representing the digits) the non-hierarchical associative relationships (lateral relationship) among the different facets and between concepts in different facets may be represented. In a thesaurus hierarchical relationships are represented by Broader Terms (BT) and Narrower Terms (NT) and lateral relationships are represented as Related Terms

(RT). Role indicators may be used to indicate the nature of relationship among concepts. S.R. Ranganathan formulated five (5) guiding principles for facet sequence and twenty-four (24) principles for sequencing the isolates; and six principles of phase relationship. Subsequent studies have identified over forty lateral relationships (LR) grouped into a dozen categories (Neelameghan and Raghavan, 2005). These principles are applicable wherever ideas have to be arranged in a helpful sequence (e.g. discourse, presentation in text, etc.)

Categorization of Concepts / Ideas: A subject in its embodied form used for communication comprises of one or more concepts connected in certain relationships to each other (concepts-net). In most cases 'time' and 'space' concepts are constituents. 'In most general schemes for subject classification (e.g. Library of Congress classification, Dewey Decimal Classification, and S.R. Ranganathan's Colon Classification) treat Time and Space as Common Elements, usually enumerated in a Common Schedule, that can be attached in the representation of a subject.

9.4. MODES OF FORMATION OF SUBJECTS AND RELATIONSHIP AMONG CONCEPTS

It is worthwhile to examine the major modes of formation of and morphing in subject domains that take place as the domain develops over a period of time and the resulting relationships among the concepts in the subject domains formed by different modes of formation. Subjects formed by the same mode are likely to display similar types of relationship among the respective concepts they are constituted out of. The modes of formation of subjects may be categorized as follows: Fission; Fusion; Distillation; Clustering; and Agglomeration (Neelameghan, 1992).

These formations are affected by interdisciplinary interactions and exchanges. Such interactions lead to collaboration, an important feature of research since about the mid-20th century, more particularly in science and technology fields. Interdisciplinary interaction may involve:

- Juxtaposition of studies on a phenomenon;

- Interdisciplinary borrowing of data, e.g. use of population data in educational planning, housing planning, etc.; representation of Hindu cosmological concepts and data in Cambodian temple architecture;
- Applications of methodologies, techniques and tools in a subject domain, of those developed in parent subject(s) or developed for other field(s) – e.g systems analysis and operations research in library studies; mathematical modeling in aerodynamics ; carbon dating applications in fossil studies;
- Application of principles and theories developed in one subject field in another subject field, e.g. astrology in future forecasting; and
- Convergence of viewpoints.

Other types of interdisciplinary interactions / connections include the following (Neelameghan and Hemalata Iyer, 2003):

Comparison studies between different subjects, e.g. comparative studies of Concept of Time in some Eastern and Western Cultures;

Differentiation studies between two or more subjects, e.g. Study of differences in the Representation of Time in Different Hindu Calendars;

Bias or **Orientation** : A document on a subject oriented for use of a different specialist group, e.g. Statistics for nurses;

Impact : Studies on the impact of the developments in a subject on other subject area(s), e.g. Impact of Information Technologies on Archaeoloy; and

Evaluation of certain concepts of a domain from the point of view of another domain, e.g. Future Forecasting by Indigenous Communities of the Americas from a Scientific Point of View.

Such interdisciplinary interactions and exchanges have been greatly facilitated by inter-country and international movement of people for study, research, participation in conferences, business etc.; enhancement of communications – telephone, mobile phones, fax, fast mails, and networking, the Internet / WWW, social networking, blogging, twittering, virtual conferencing etc. New concepts, new fields of study and new technologies are emerging much faster than ever before. The time-lapse between the emergence of an idea / theory and its application (e.g. to create new devices) has been reduced progressively (*see* Table 1).

Table 1: Reduction in Time-lag between Theory formulation and its Practical Application

Subject	Year of Principle formulation	Year of Application	Number of Years (lapse)
Photography	1727	1839	112
Telephone	1823	1876	53
Atomic power release	1932	1945	13
Transistor	1940	1948	8
Laser	1958	1960	2

The recent past saw the fast emergence of such interdisciplinary subjects as biotechnology, computational linguistics, bio-linguistics, socio-linguistics, medical cybernetics, nano-science and nanotechnology and so on, although their roots may be in subjects that have been researched for decades or even centuries. These interdisciplinary formations emerge over a period of time. An implication is that KOTs should be dynamic and quickly updatable to accommodate the emerging ideas in science, technology, and new devices, products, services, etc. (Neelameghan, 2000).

Fission is (internal) division into smaller units as in cell fission (natural or artificially induced), and nuclear fission. The former may be relatively slower compared to the latter. The methods and techniques of fission may come from other subject areas. In the case of Time such smaller units (e.g. seconds, minutes, hours, days, months, years, centuries, eons, etc. are widely recognized) and provided for in almost all KOTs, usually in a hierarchy (hierarchical relationship). This mode is perhaps the earliest.

Fusion: The application of theory, principle, techniques, etc.

borrowed from other discipline(s), may give rise to an Application relationship between the host discipline (the field applied) and the discipline from which they have been borrowed. Over a period of time there may be integrations of concepts or of whole domains and new concepts and / or subjects may arise. Associating them or placing with one or the other - host or donor - subject may not be appropriate or helpful to the specialist users. New societies, new periodicals, new courses of study, etc. may concretize over time the new area with a new name. The mode of formation of the new subject is by Fusion, that is, the integration of concepts of two or more subjects that are mutually exclusive, independent disciplines, resulting in the creation of a new discipline. Examples: astro-chemistry, astro-biology, biotechnology, computational linguistics, bio-linguistics, socio-linguistics, and medical cybernetics.

Distillation: A particular idea or group of ideas falling within a particular domain studied in great depth and detail giving rise to considerable amount of literary warrant about these concepts. New principles for guiding research in this area may be formulated and eventually there may be less and less frequency of reference to or use of the theories and normative principles of host subject domain. A stage may be reached when studies on the particular concept(s) may become a field of specialization by itself – that is the concepts becoming the core. Again, the starting of new courses of study, periodicals and research groups devoted predominantly to studies on these concepts will signal the formation of a new area of specialization. The mode of formation is Distillation. Examples: Operations research, Information theory, Cybernetics, Microbiology, Geodesy, Applied linguistics, and Time studies

Clustering: A collection of documents dealing with

- an Area or a Country (e.g. China: Developments in science, technology, agriculture, fauna and flora, geological studies, languages and literature, crafts and fine arts, philosophy, religion, sociological studies, education in China, history, politics, economy, legal system). The collection may be called Sinology. Similarly for other areas / countries;

- an entity (e.g. Time). The concept may occur with any subject. But a document examining Time from different points of view may not be associated with one subject domain (*see* below)
- a person who has written or writes and discourses about many different subjects and whose life and work are viewed from different (subject) perspectives. In such a case the person's biography, collection of papers by and on the person may not be placed under or associated with any one subject domain (e.g. Mahatma Gandhi)

The Time element being ubiquitous may occur in association with almost any subject domain. It means it is a common concept. In this case most schemes for subject classification provide a schedule of common Time isolates (cf. DDC, LC, CC).

Time may be studied in the context of a particular discipline, e.g Physics. In this case, subject classification schemes may treat 'Time' as a fundamental element along with Space and Energy.

There are studies in which the concept 'Time' occurs as a qualifier to another concept. For example: Early morning sun; Morning sun; Evening sun; Twilight sun; Evening melodies; effects of working in night shifts; time-lapse photography; time-balls in signaling devices; time capsule; time card; time lock; timepiece; timetable; World War 1939-1945; Hundred years War. In CC such a qualifier concept is named a Speciator. Suffixing a Speciator to a concept gives rise to hierarchical divisions. Example:

pann (raga)

- *pakalpann* (day ragas)
- *iravupann* (night ragas)
- *potupann* (all-time ragas)

For further details *see* chapter G.

There are concepts in which 'Time' is 'qualified' by other concepts. Example:

Name of Concept	Representation in CC
Opportune Time	Time-opportune
Leisure Time	Time-Leisure
Work Time	Time-Work
Wasted Time	Time-Wasted
Meal Time	Time-Meal
Breakfast Time	Time-Breakfast
Dinner Time	Time-Dinner
Supper Time	Time-Supper

There are also studies in which 'Time' is a facet. For example Martin Clayton's book *Time in Indian Music* (2008). Here 'Time' occurs as an Attribute, a Property of music.

Time may be studied from different perspectives, such as the present book. Another example: Time and the sciences / edited by Josefina Mena Abraham and Frank Greenaway; Unesco, 1979. Information seekers on "Time studies" or "Time viewed from different perspectives" will find it more helpful if all such studies on Time are brought together in a single cluster. CC provides for such clusters in the Class "Entity Studies". An alternative is in the alphabetical subject index, KWIC, KWAC, PRECIS, POPSI.

9.5. SYSTEM AND STYLE

In many disciplines the concept of 'system', the framework, the basis, is either implicit or explicitly mentioned. For example: Euclidean geometry; Physics usually meaning Newtonian physics, Quantum physics; In Medicine: Ayurveda, Allopathy, Unani, Homeopathy; In Psychology: Experimental psychology, Gestalt psychology; In education: Montessori, Rousseau, Kant, Pestalozzi; In economics: Capitalism, Socialism, Communism, Technocracy; In jurisprudence: Socratic School. Parallel to "System", in Literature and the Arts the framework may be "Style", denoting a particular period. For example Elizabethan poetry, Tamil classical literature, Moghul paintings, Renaissance architecture. In subject classification schemes, the system or style is indicated by the year, decade, century the particular system was propounded or became prominent. For the favoured system (in relation to a particular library or information system) such an indication may be omitted.

Cultural Impact : To help overcome confusion that may be caused by socio-cultural differences in expressing time is the need for ontology of SCTE. Wennerberg and Schulz (2010, p.64) state that "the purpose of the ontology is to explicitly and formally describe these context dependent temporal expressions. This is done by defining appropriate hierarchical categories to which the expressions will be assigned. The ultimate goal is to be able to develop context aware temporal applications for the Web such as calendars or appointment schedulers".

9.6. SEMANTIC RELATIONSHIPS IN CULTURE-SPECIFIC DOMAINS

Conceptual relationships between terms in the Humanities are not as precisely defined as in the physical or life sciences. In addition to this general issue, building a multilingual thesaurus for a culture-specific domain raises certain issues that are related to the nature of such domains. Firstly, concepts encountered in and associated with culture-specific domains in particular are abstract in nature and rarely can they be related to concrete referents. Secondly, a large number of concepts of a culture-specific domain are those that have some meaning in the life of the members of the community belonging to the culture. These have implications for a multilingual thesaurus. A language is a product of, and reflects the culture of the particular community. It is the culture and lifestyle prevalent among the members of a particular community that necessitates and results in the formation of lexemes / expressions (words / terms) for concepts associated with that culture and lifestyle. It is therefore very likely that, unless the communities that speak two different languages share the same culture, certain concepts in culture-specific domains may have verbal expressions only in a particular language. The following case study drawn from the construction of Tamil English thesaurus brings out some of the issues involved and how they were resolved.

In building a Tamil English thesaurus it was found difficult to find exact equivalent concepts in the English language for Tamil terms. The notion of '*equivalence relationship*' widely used in information retrieval thesauri had to be extended to include both equivalent terms in the source language (synonyms in Tamil for Tamil terms) and terms in English denoting near-equivalent / similar concepts. The second major issue, also related to the notion of '*equivalence relationship*', was the occurrence of several different terms in Tamil with closely similar meanings / connotations. The problem was how to handle this situation and to understand the implications for the bilingual thesaurus. By examining two synonymous words, '*tAmarai*' and '*kamalam*' and their English equivalent, "lotus" retrieved from the online Tamil Lexicon http://dsal.uchicago.edu/ dictionaries/tamil-lex/) it was found that the following major issues needed attention in building the bilingual thesaurus:

- There appears to be a large number of synonyms in Tamil (this is probably true of other Indian languages also). Synonyms in the form of variant forms of spelling also appear to be in wide use in the Tamil language;
- A large number of RTs (laterally related terms) also appear to characterize Tamil language. For most of the terms related to a descriptor, there was no single equivalent word / term in English. This may suggest the need for using phrases / compound terms in English to denote many of the concepts encountered; and
- Many of the terms that are related to '*tamarai*' (as synonyms, NTs and RTs) were found to be homographs with multiple connotations. This creates problems especially if a related term is used as the descriptor to denote some other concept. The importance of employing appropriate role operators (individualizing elements) to clearly indicate the connotation of a term cannot be over emphasized.

The large number of RT terms to '*tamarai*' may be grouped as representing broad ideas / concepts that are associated with lotus, such as a plant, a flower, shape of a lotus, and attributes like tenderness of lotus petals.)

The issues raised are representative of the problems common to most Indian languages. These issues need to be addressed and appropriate guidelines developed. Besides, there are also issues related to transliteration, etc.

Metaphorical Semantic Representation : Ontological study of words and names provide important insights for lexicographers and builders of thesauri. This is explained by taking the example of the Tamil term '*kuRinci*' and kurinji lore in the context of the naming practices in early Tamil and their social and cultural associations (Neelameghan and Raghavan, 2012). The term *kuRrinci* denotes a specific species of flower- henna, cone head; it also refers to the type of land- hilly tract; tells an association with the type of melody of kurinji region; also means poem- *kurinci-p-pattu*, and a connection with clandestine union of lovers associated with the hilly tract. The usage clearly shows the association of social and cultural connections of the word and its metaphorical connotations. For lexicographers and thesaurus builders the issue is how to bring out various relationships, particularly those concepts having metaphorical connotations. In lexicons contextual variations in meaning is shown immediately after the word, whereas in a thesaurus it can only be brought out by annotations, which is not a practical solution when metaphors are concerned. There is a time/age connection between kurinji lore and the tribal lives of Muthuvas of Munnar and the Todas of Nilgiris. They count their age as the multiples of the flowering cycle of kurinji. For these tribes kurinji symbolizes divinity, love and romance.

The Sanskrit word '*akshara*' is good example of a word having denotative, connotative and metaphorical usage as shown below:

- Alphabet in Sanskrit language (denotative);
- Imperishable, non-destructive, non-decaying (connotative and temporal);
- Brahman, Siva, Vishnu (metaphorical and temporal); and

- Om meaning highest Brahman and also Pranava (denotative and synonymous)

This evolution of names is intimately entwined with the natural geography (flora and fauna) and time horizon of the culture. A similar evolution is also found in some of the indigenous people of North America (gchester@charter.net).

Homonyms and Homographs : In devising mechanisms for storing and retrieving information, whether in indexing or thesaurus construction, the deviser has to take care of the problems caused by homonyms, homographs, homophones, and synonyms. This is generally done by providing suitable cross references or annotations keeping in mind the cultural and linguistic pattern in which the domain knowledge evolved. For instance, with பதினெண்கீழ்க்கணக்கு as search term will retrieve the related entry/entries in the Tamil-English thesaurus hyper linked to the online University of Chicago Tamil Lexicon (Dictionaries of South Asian Languages, http://dsal.uchicago.edu/dictionaries/tamil-lex/) displaying the different meanings of the term along with the derivation, contextual occurrence, etc., all brought together under one entry (*See also* Neelameghan and Raghavan, 2009). Another problem is dealing with disambiguation caused by one and the same term (spelt and/or pronounced the same way) may occur in different contexts and have different meanings.

Literary Devices in Scholarly Communication : Literary devices or structural and thematic devices are used by authors to aid the reader / listener in better understanding his / her work – composition and exposition – and the relationship of the work to the work/works of other author/authors. These devices are also intended to make the exposition transparent to different categories of readers. To help scholarly communication guidelines for authors and good writing guidelines are available. It is interesting to note that predating Christian era the Tamil work *TolkAppiam* a comprehensive treatise on linguistics and poetics embodying an organized corpus of norms, conventions and standards that define and describe the linguistic usage and the theory of literature was composed by TolkAppiar. It also provided standards and guidelines for various types of expositions in practice at that time. An example is the similarity between 'modern' guidelines and those prescribed in the Tamil grammar classic *Tolkappiam* for literary composition, in particular the use of literary devices (Neelameghan and Parthasarathy, 2008). Indexing and retrieval devices, such as a thesaurus, can aid research in the subject / domain covered by the thesaurus.

Ontological Categories and Semantic Relationships. To observe semantic relationship between concepts ontological categorization is a necessity. This can be seen by studying the Ontological Categories *attikAyam* or *astikayam* that arose while developing the Tamil-English bilingual thesaurus for classical Tamil studies. From the broad category (BT) *paNcAttikAyam* – "under which 'reals' are distributed" further sub-categories are derived. One such sub-category is NT *jIvAstikayam.*

NT *jIvAstikayam* (*jIv* or *Atman* representing conscious-sentient principle) is further categorized into:

(a) *Taram Astik Ayam*, i.e. the entity enabled by *j...v* to reach its goal (RT) and

(b) *ataram AstikAyam* or that "which without any movement on its part attracts and keeps passive the thing attracted". Example, "the shade (shadow) of a tree first attracts a person walking along the road on a hot day and then keeps the person quietly resting under it". The principle is also used in other contexts like a devotee is attracted to god (tree in the example) for protection or salvation. Similarly, other categories (NT) are established.

The principle and relationship in a) above may be generally expressed as Entity- (through enabler) - RT Goal/Destination, and b) above as Entity X – (attracts)- RT Entity Y / Entity- attracted to RT Entity X.

9.7. MULTILINGUAL ISSUES

Some of the above-mentioned aspects of Time may be

applicable to information systems handling records in two or more languages. For better results *vis a vis* satisfying users' information needs two or more KOTs may be used simultaneously or integratively. Neelameghan and Raghavan (2012) have discussed the problems that arise in handling multi-lingual records.

Epilogue

Information systems hold mainly embodied knowledge and their surrogates in document, non-document, and /or digital form. A variety of means and methods and tools are used for organizing, searching and retrieving the embodied knowledge. We therefore began with brief notes on knowledge, and modes of valid knowledge. The concept of time is also elaborated in the first chapter. Certain expressions of time in information studies not influenced by cultural differences are discussed as also how these differences impact expressions of time in information systems.

The embodied knowledge in all the domains is impacted by cultural differences and this will be reflected in the knowledge organization and its tools. Hence, the subject of this book is multi-focal - temporality, social and cultural knowledge. We discussed the differences in the concept, perception, experience and representation of 'Time' in different cultures and the impact of such differences in cultural frames on knowledge and knowledge organization.. We discussed the impact of the differences on the expressions of time in different domains, more particularly in the humanities – literature, arts and fine arts, evidence-based studies, information science and systems and services, and on knowledge organization tools. All these facets are interlinked as they are the characteristics of one common entity, Time, and they influence each other.

In recent times there has been a perceptible increase in the number of writings on multicultural issues *vis a vis* knowledge organization. Papers contributed to the ISKO (International Society for Knowledge Organization) conferences and to its journal *Knowledge Organization* reflect this trend. Smiraglia and Hur-Li Lee editors of *Cultural Frames of Knowledge* (Ergon Verlag, 2012), in their concluding essay "Conclusions: Toward multicultural Domain Plurality in Knowledge Organization" analyzing the papers contributed to the volume also point out this trend.

In their paper "Frames of Knowledge" Neelameghan and Raghavan (2012), write:

> *"We live in a culturally diverse world. The nature of concepts and knowledge structures, particularly in the culture-specific domains are functions of the socio-cultural framework characterizing a society. The overall cultural traditions, beliefs, customs, language, literature, art forms, etc. may be seen as a frame within which to interpret and understand the culture. This frame is not static, as in the global perspective there are many cultures, traditions and philosophies among which interactions and exchanges take place, leading to modification of existing frames and/or creation of new ones."*

This perspective applies to the concept and measurement of 'Time' as discussed in this book. Despite the differences, however, there are several areas which present unity of concepts. More of such unity / correspondences that run across cultures, traditions and philosophies need to be further explored. It will lead to better and proper understanding of the cultures and practices. In this context, information science and its knowledge organization tools have important roles to play.

References

Abou-Zeid, E (2003). Towards a cultural ontology for interorganizational knowledge processes, In: *Proceedings of the 36th Hawaii International Conference on System Sciences HICSS-36*, pp. 4-9. 2003

Adam, B. (1990). *Time and social theory.* Cambridge: Polity Press.

Adam, B. (1995). *Time watch: The social analysis of time.* Cambridge: Polity Press

Agnesund, M. (1997). Representing culture-specific knowledge in a multilingual ontology –an object-oriented approach. *Proceedings of the workshop on ontologies and multilingual NLP, Nagayo, Japan.* http://citeseer.nj.nec.com/agnesund97representing.html.

Akhtar, Shahid, Ed. (1993). Equity, growth and participation: the information age in Development, *Journal of the Society for International Development (Italy)*, 3, 1-86 (whole issue). Special issue on Communication and Development.

Albertazzi, Liliana (1996). Formal and material ontology. In: *Poli, Roberto and Simons, Peter, Ed. Formal Ontology.* Kluwer 1996; p. 199.

Allen, J. et. al. (2002). A DAMAL ontology of time. *Tech. Report*, DAML, org.

Arrow, Kenneth (1996). The economics of information: an exposition. *Empirica*, 23(2); p.119-128.

Atran, Scott. Itzaj Maya folkbiological taxonomy: Cognitive universals and cultural particulars. In: *Folkbiology,* edited by Douglas L. Medin and Scott Atran. MIT; 1999. Chapter 6.

Avramescu, A. (1975). Modeling scientific information transfer, *International Forum on Information and Documentation,* 1(1); p. 13-19.

Bates, Marcia J. (2006). Fundamental forms of information. *Journal of the American Society for Information Science,* 57(8); p.1033-1045.

Bell, Daniel (1976). *The coming of postindustrial society: A venture in social forecasting.* Ed.2. New York, Basic Books

Berman, Francine; Hey, Anthony J. G. and Fox, Geoffrey C. (2003). *Grid computing: making the global infrastructure a reality.* Wiley. http://www.grid2002. org/.

Bertalanffy, Ludwig von (1968). General systems theory - a critical review, In : *Modern Systems Research for the Behavioral Scientist,* p. 11-30 / Edited. By Walter Buckley, Chicago, Aldine Publishing Co.

Beule, J. De. (2003). Creating temporal categories for an ontology of time. *Proc. of CLIN 2003, 14th Meeting of Computational Linguistics in the Netherlands,* 19 December, 2003, Antwerpen, Belgium

Boulding, Kenneth E. (1968). General systems - the skeleton of science, In; Buckley, Walter, Ed. *Modern Systems Research for the Behavioral Scientist,* p. 3-30, Chicago, Aldine Publishing Co.

Bradford, S.C. (1948). *Documentation.* London, Crosby Lockwood.

Burton, R.E. and Kebier, R.W. (1960). The "half-life" of some scientific and technical literatures, *American Documentation,* 11(1); p. 18-22 http://onlinelibrary.wiley.com/doi/10.1002/asi.5090110105/abstract

Business Standard (2011). Cutting language barriers. New language translation tool unveiled at the 20th International WWW

Conference. *Business Standard, Technology,* March 31. p.10.

Castells, Manuel (2000). The rise of the network society. Ed. 2. Oxford (UK), Blackwell Publishers; v. 1

Chevalier, Jacques, M. and Buckles, Daniel J. (2008). SAS: A guide to collaborative inquiry and social engagement. International Development Research Centre, Ottawa. New Delhi, Sage Publications; p.11

Chugh, Manoj (2010). Where will you store Zetabytes of information? *Business Standard,* 15 July; p.11.

Circadian rhythm. http://en.wikipedia.org/wiki/Circadian _rhythm

Clayton, Martin (2008). *Time in Indian music: Rhythm, metre, and form in North Indian rag performance.* Paper back. Oxford University Press.

Concept of time in different cultures. http://www.library thing.com/topic/56455. Accessed 6 June 2011.

Corazzon, Raul (2010). *Theory and history of ontology.* http://www. Formal ontology-definitions-one.httm Accessed 19th July, 2010.

Crystal, David (2005). *The language revolution.* Cambridge: Polity Press, 2004; Indian reprint 2005.

Dalrymple, William (2009). *Nine lives: In search of the sacred in modern India.* London: Bloomsbury.

Data, data everywhere (2010). A special report on managing information. *The Economist,* February 27th, p.1-16.

Davies, Elizabeth (2000). *Unlocking the secret sounds of language: Life without time or numbers.* www.indipendent.co.uk/news/science/ulocking-the secret-sounds-of-language-life-without-time-or-numbers

De Bruijn, B. (1999). Foundations of demographic theory. *PDOD Publication Series A. Doctoral dissertation.* The Netherlands: University of Groningen

De Solla Price, Derek (1963). *Little science, big science,* New York: Columbia University Press. (The George B. Pegram Lectures)

Doerr, M., Le Boeuf, P., and Bekiari, Ch. (2008). FRBROO, a conceptual model for performing arts. In *Conference proceedings "The Digital Curation of Cultural Heritage" ICOM-CIDOC Annual Meeting. 15-18 September*, Athens, Greece

Durett, Rick (2010). Some features of the spread of epidemics and information on a random graph. *Proceedings of the National Academy of Sciences of the United States of America*, 107(10); p. 4491-4498. http://www.pnas.org/content/107/10/4191.shprt

Dutt, Devina (2012. A time for music. *The Hindu Magazine*, September 2: p.8

Elder, G.H.J. (1974a). *Children of the great depression*. Chicago, Ill.: Chicago University Press

Elder, G.H.J. (1974b). Time, human agency and social change: Perspectives on the life course. *Social Psychology Quarterly*, 57(1): p.4-15

Elmore, Garland C. (2001). *Improving teaching and learning with technology: Perspectives on change and faculty engagement*. www.statewideit.iu.edu/sitc_sites/statewide01/ tatewide01/ppt/elmore.ppt

Everett, Daniel. *Piraha people*. Wikipedia: en.wikipedia.org/wiki/Piraha-people

Evers, Hans-Dieter and Gerke, Solvay (2003). Local and global knowledge: Social science research on southeast Asia. Paper read at an International Conference *Social Science in a Globalising World: Contemporary Issues in Asian Social Transformation", UNIMAS Kuching 22-23 Sept 2003*. (Department of Southeast Asian Studies, University of Bonn, Bonn).

Evica, George M. (1963). Time's arrow and the world's order: a re-evaluation of bonding theory, *ETC: A Review of General Semantics*, 20(4); p. 421-37.

Fairthorne, Robert A. (1969, 2005). Empirical hyperbolic distributions

(Bradford-Zipf-Mandelbrot) for bibliometric description and prediction. *Journal of Documentation*, 61(2); p.171-193, 2005. (Republished as 60 Years of the Best in Information Research Progress in Documentation). Originally published in 1969.

Feynman, Richard (1963). *Lectures in physics*. v.1. Addison-Wesley.

Foucault, M. (1977) *Discipline and punish*. London: Allen Lane, p. 27.

Ganesh, K.N. (2011). Space-time, event and expression in early Tamil texts: Historicizing Tolkappiyam. *Indian Historical Review*, 38 (10) p.1-22.

Gerth, Hans Heinrich and Mills, Charles Wright (1958). *From Max Weber: essays in sociology*. p. 215. Oxford University press.

Giddens, A. (1981). Agency, institution and time-space analysis. In: *Advances in social theory: Toward an integration of micro- and macro-sociologies*, ed. by Knorr-Cetina and Cicourel; p. 161-174. London: Routledge & Kegan Paul

Giddens, A. (1984). *The constitution of society: Outline of the theory of structuralization*. Cambridge: Polity Press.

Giddens, A. (1991). *Modernity and self-identity: self and society in the late modern age*. Cambridge: Polity Press.

Giddens, A. (1994). Living in a post-traditional society. In: *Reflexive modernization*, p. 56-109. Cambridge: Polity Press

Goffman, W. (1966). A mathematical approach to the spread of scientific ideas. *Nature*, 212; p.449-452.

Goffman, W. and Newill, V.A. (1964). Generalization of epidemic theory; an application to the transmission of ideas. *Nature*, 204; p.225-228.

Goffman, W. and Newill, V.A. (1967). Communication and the epidemic processes. *Proceedings of the Royal Society, Series A*, 298; p.316-334.

Goffman, W. and Warren, K.S. (1969). Dispersion of papers among journals, based on a mathematical analysis of two diverse medical literatures. *Nature*, 221; p.1505-07.

Gruber, Thomas (1993). A translation approach to portable ontology specifications, *Knowledge Acquisition*, 5; p. 199-220

Gruber, Thomas (2007). Ontology of folksonomy: a mash-up of apples and oranges. *International Journal on Semantic Web and Information Systems*, 3(2). (Originally published on the web in 2005)

Gruber, Thomas (2008). Collective knowledge systems: where the social web meets the semantic web, *Journal of Web Semantics*, 6(19); p.4-13. tomgruber.org/writing/index/htm.

Gutterman, Alan S. (2011). *Kluckholm and Strodtbeck's cultural value orientation*. http://alangutterman.typepad.com/files/cms-Kluckholm-strodtbeck-cultural-dimensions.pdf/

Hall, Edward T. (2011). *Monochronic and polychromic time*, Chapter 5, p.3131-319

Harshananda, Swami (2008). *A concise encyclopaedia of Hinduism*, 2; p. 317-318. Bangalore: Ramakrishna Math. 3 v.

Hartman, L. M. (1966). *The prospect of forecasting technology*, p. 237, New York, Macmillan Co.

Hawkins, Donald T. (2001). Information science abstracts: tracking the literature of information science. Part 1: Definition and map. *J. American Society for Information Science and Technology*, 52(1); p.44-53.

Hegde, B.M. (2011). Intellectual terrorism. *Bhavan's Journal*, 58(2); p.59.

Hirshleifer, J. (1971). Private social value of information and the reward to inventive activity, *American Economic Review*, 61; p. 561-574.

Hogan, D.P. (1981). *Transitions and social change: The early lives of American men*. New York: Academic Press.

Hollingsworth, T. Deirdre, Ferguson, Neil and Anderson, Roy M. (2007). Frequent travelers and rate of spread of epidemics. *Emerging Infectious Diseases*, 13(9). Sept.http:// www.cdc.gov/EID/content/13/9/1288.htm http://zdnet.com/blog/btl/

2010-data-doubling-every-11-hours/4497. (accessed 5th September 2010).

http://en. Wikipedia.org/wiki/sociocultural_evolution, accessed 15 March 2011

http://family.jrank.org/pages/1074/Life-Course-Theory.html# ixzz1vNLEjUFs)

http://library.thinkquest.org/3493/noframes/chaos.html, accessed 11 March 2011.

Illich, Ivan (1975). *Limits to medicine*. London: Calder and Boyars

Isenson, R. S. (1966). Technological forecasting in perspective, *Management Science*, (Providence, R. I.), p. 370-383.

Jenkins, Orville Boyd (2003). *Time – a cultural concept*. http:// orvillejenkins.com/acosscultures/ time in cultures.html (accessed 5 June 2011)

Jenkins, Orville Boyd (2008, 2009, 2010). *Time in different cultures*. http://orvillejenkins. com/acosscultures/time in cultures.html (accessed 5 June 2011) jharkhand.us/culture.aspx. Visited July 2011.

Jorna, Kerstin and Davis, Sylvie (2001). Multilingual thesauri for the modern world – no ideal solution. *Journal of Documentation*, 57(20;p.284-295.

Kak, Subbash (2000), *The astronomical code of the Rg Veda*. New Delhi: Munshiram Manorharlal Publishers Pvt. Ltd.

Kohli-Khandekar, Vanita (2011). Using Wikipedia to change the language of the Web. *The Hindu*, January 5 Opportunities, p.4.

Korzybski, Alfred (1924). Time-binding: the general theory, presented before *The International Mathematical Congress*, Toronto, 1924.

Koslu, Kasla, Trancy, Cedric and Jorg Golzhauser, Jorg (2003), www.eurotimes.ae.krakow.pl/ files/el1(5).2005/kktjg_limes1 (5).pdf)

Krishnaswamy, M.H. *Concept of time in Hindu mythology*. http:// www.templenet.com/ beliefs/concept_of_time.htm

(Accessed July 2011)

Kurup, Deepa (2011). Using Wikipedia to change the language of the Web. *The Hindu,* January 5 Opportunities, p.4

Le Lionnais, Francois (1960). *The Orion book of time.* English translation edition. New York, N.Y.: Orion

Landauer, Rolf (1991). Information is physical. *Physics Today,* 44(5): p.23-29

Life course – http://family.jrank.org/pages/1074/Life-Course-Theory.html#ixzzlvNLEjUFs (visited 20 May 2012)

Lloyd, Seth and Ng, Y. Jack (2004). Black hole computers. *Scientific American,* 291(5): p.30-39

Lorenz, Edward N (1972). *Chaos and fractals – the butterfly effect.* http:www.stscl.edu/ ~bradley//seminar/butterfly.html

Luhmann, N. (1977; 1982). The future cannot begin: Temporal structures in modern society. In: *The differentiation of society,* p.271-288; New York: Columbia University Press

McGarry, Kevin, J. (1975). *Communication, knowledge and the librarian,* p. 46, London, Clive-Bingley.

MacGill, Victor (2007). *A non-technical introduction to the new science of chaos and complexity.* Ed Lorenz and the butterfly effect. http://Complexity,orconhosting,net.nz/ butterfly.html

Machlup, Fritz (1980). *Knowledge: Its creation, distribution, and economic significance, v. 1: Knowledge and knowledge production.* Princeton, Princeton University Press; p. 108-109.

McWhorter, John. (2009). *Our magnificent bastard tongue - the untold history of English.* Audible Inc..

Manmadhan, Prema (2011). Rhythmic pangs of the past. *The Hindu,* Metroplus June 4.

Marschak, Jacob. (1968). Economics of inquiring, communicating, deciding. (Richard T. Eli Lecture). *American Economic Review,* 58(2); p.1-18.

Martin, William J. (1995). *The global information society.* London, Aslib

Mead, G.H. (1932; 1980). *The philosophy of the present.* Chicago, Ill: Chicago University Press.

Medin, Douglas L. and Rips, Lance J. (2005). *Concepts and categories: Memory, meaning and metaphysics,* p. 37-72. In Holyoak, Keith J. and Morrison, Robert G. Ed. Cambridge, Mass: Cambridge University Press.

Mehta, Surendralal G. (2011). Space, time and us (From the President). *Bhavan's Journal,* 58(5); p.12-13

Merriam-Webster online, http://www.merriam-webster.com/dictionary

Mills, Melinda (2000). Providing space for time: the impact of temporality on life course research. *Time & Society,* 9(10); p.91-127.

Mito, Yoko (2007). Corporate culture as a strong driving force for punctuality – Another 'Just in time'. *Hitachi-Rail.com* (25 March 2007)

Mohanty, J.N. (2002). The concept of intuition. In: *The bases of Indian culture:commemoration volume of Swami Abhedananda.* Ed. by Amiya Kumar

Morphet, C. S. (1974). A probabilistic model of time-dependent factors governing the acceptance of innovative ideas by industry', *R & D Management,* 4(3); p. 165-75.

Mukherjee, S., Bhattacharya, B.N. and Singh, K.K. (1996) Distribution of time of first birth in presence of social customs regulating physical separation and coital frequency. *Mathematical Biosciences,* 13(1):1-21

Munshi, R.N. (n.d.) *World famous quotations.* New Delhi: Goodwill Publishing House.

Naisbitt, John (1982). *Mega trends: ten new directions transforming our lives.* Warner books.

Naisbitt, John (1990). *Megatrends 2000: ten new directions for the 1990s.* William Morrow & Co. Inc.

Nakamura, Hajime (1993). *The notion of time in India* (summary by

Angie Weicekauskas, 1993) http://www.drury.edu/ess/Culture/India.htm (visited 1 September 2011)

Narayana, G.J. (2010). *Knowledge and information: perspectives and prospects*, p.147-149. New Delhi, Ess Ess Publications

Neelameghan, A. (1973). Social change, communication of ideas, and library service with special reference to developing countries, *Library Science with a Slant to Documentation*, (Bangalore), 10(1); p. 1-29.

Neelameghan, A. (1979). Expressions of time in information science and their implications. In: *Time and the sciences*. Ed. by Josefina Mena Abraham and Frank Greenaway Paris: UNESCO; 1979; p. 103-117

Neelameghan, A. (1992). Ranganathan's generalized model of subject structure and modes of formation of formation of subjects. In: *Cognitive paradigms in knowledge organization. Second International ISKO Conference*, Madras, 26-28, August 1992; p.269-285.

Neelameghan, A. (2000). Dynamism and stability in knowledge organization tools: S.R.Ranganathan's contributions. In: *Dynamism and stability in knowledge organization. Proceedings of the Sixth International ISKO Conference*, 10-13 July 2000, Toronto, Canada . Ed. by Claire Beghtol, Lynn C. Howarth, Nancy J. Williamson. Ergon Verlag.

Neelameghan, A. and Hemalata Iyer (2003). Information organization to assist knowledge discovery: case studies with non-bibliographic databases. *CCQ Cataloging & Classification Quarterly*, v37(1/2); p.115-130.

Neelameghan, A. and Parthasarathy, S. (2008). Literary devices in scholarly communication. *Information Studies*, 14(1); p.15-38.

Neelameghan, A. and Raghavan, K.S. (2006). Semantics of relationships in knowledge organization: Lateral relationship. *SRELS Journal of Information Management*, 42(4); p.361-382; Update 1, *Information Studies*, 12(2);2006;p.115-21

Neelameghan, A. and Raghavan, K.S. (2007). Online bilingual thesaurus for subjects in the humanities. *International Conference on Semantics & Digital Libraries, ICSD 2007*, 21-23 February 2007, Bangalore. Ed. by A.R.D.Prasad. and Devika P. Madalli; p.489-505

Neelameghan, A. and Raghavan, K.S. (2009). A knowledge organizing system for humanistic disciplines with enhanced capabilities: Case studies. *Information Studies*, 15(2);p.75-94.

Neelameghan, A. and Raghavan, K.S. (2012). Concept of 'time': semantic relationships and cultural frames. In: *Categories, contexts and relations in knowledge organization. Proceedings of the twelfth International ISKO Conference*, 6-9 August 2012, Mysore, India. Ed by A.Neelameghan and K.S.Raghavan. p.212-219. Wursburg, De.:Ergon Verlag.

Neelameghan, A. and Raghavan, K.S. (2012). Frames of knowledge: A perspective of Vedic-Hinduism and Dravidian culture. In: *Cultural frames of knowledge*. Ed. by Richard P.Smiraglia andHur-Li Lee. Wursburg, De.: Ergon Verlag. p. 19-62.

Neelameghan, A. and Seetharama, S. (1975). *Information transfer: the next twenty-five years: librarianship and information science in 2001*, Seminar arranged by the British High Commission, British Council Library, on the occasion of its Silver Jubilee, Madras, 7-10 October 1975; p. 9-37.(unpublished)

Neelameghan, A, Buche, V.V. and Gupta, B.S.S. (1970). Distribution pattern of the duplication of discoveries, *Nature*, 227; p.1060

Neelameghan, A, Buche, V.V. and Gupta, B.S.S (1970a). Discovery and duplication of antibiotics: a correlation. *Current Science*, 34; p.511-512

Newman, M.E.J. (2002). Spread of epidemic disease on networks. *Physical Rev.* E66, 016128 (11 pages). http:/pre.aps.prg/bstract/PRE/v66/11/e 016128

Palhegyi, Krisztina (2006). *Time and cultures.* hunReal.com. Under the Frog

Panchapakesa Iyer, A.S. and Sankara, Angarai (1989). *Karnataka sangeeta sastra: Theory of carnatic music.* Trans. into English by T.T.Vijayaraghavan. Madras: Ganamruta Prachuram.

Porat, Marc and Rubin, M. (1977). *The information economy: definition and measurements.* Washington, D.C., U.S.Office of Telecommunications, U.S.Department of Commerce.

Pritchett, Price (1998). http://www.google.co.in/#hl=en&source=hp&biw=1021&bih=593&q= Price+Pritchet+5000+years&btnG=Google+Search&aq=f&aqi =&aql=&oq=Price+Pritchet+5000+years&gs_rfai=&fp=e791fe993fa412ba

Raghavan, K.S.(2010). South Asia Support System for Greenstone: a report. *Information Studies,* 16(4); p.201-213

Raju, Anupama (2009). Writers without borders. *The Hindu, Literary Supplement,* December 6, p.1

Ramakrishna Rao, K. (2010). *Yoga and parapsychology: Empirical research and theoretical essays,* p.1-41. Delhi: Motilal Bnarsidass Publishers.

Ranganathan, S.R. (1931). *The five laws of library science.* Madras: Madras Library Association.

Ranganathan, S.R. (1960; reprint 2006). *Colon classification,* 6th ed. Chapter 7.New Delhi: Ess Ess Publications for Sarada Ranganathan Endowment for Library Science

Ranganathan, S.R. (1967; 2006). *Prolegomena to library classification.* Ed.3. Assist. by M..A. Gopinath. Delhi: Ess Ess Publications.

Ranganathan, S.R. (1967; reprint 2009). *Ramanujan: the man and the mathematician,* p.25. B5. New Delhi: Ess Ess Publications for Bangalore: Sarada Ranganathan Endowment for Library Science

Ranganathan, S.R. (1973; reprint 2006). *Documentation: genesis and development,* p.48-50. New Delhi: Ess Ess Publications for Bangalore: Sarada Ranganathan Endowment for Library Science.

Ranganathananda, Swami (2008). Message of the Upanisads. In: *Swami Ranganathananda: dynamic spirituality for a globalized world. A commemorative volume of selections from his books*, Chapter 9, p. 149. Hyderabad, India: The Citizens' Committee, Birth Centenary Celebrations of Swami Ranganathananda.

Ravichandra Rao, I.K. (2010). *Growth of literature and measures of scientific productivity: scientometric models.* New Delhi: Ess Ess Publications; Bangalore: Sarada Ranganathan Endowment for Library Science.

Rifkin, Jeremy (1998). *The biotech century: harnessing the and remaking the world.* New York, Jeremy P. Tarcher (Putnam).

Rogers, Everett M. and Stanfield, J. David (1968). Adoption and diffusion of new products: emerging generalization and hypothesis, In: *Applications of the sciences in marketing management* / Frank M. Bass, Charles W. King and Edgar A. Pessemier, p. 227-250, New York: John Wiley & Sons, Inc.

Sambamoorthy, P. (1983). *South Indian music Book 3*, 8th ed. Madras: Indian Music Publishing House.

Sambamoorthy, P. (1987). *South Indian music Book 2*, Madras: Indian Music Publishing House.

Satprakashananda, Swami (1974; 4th Indian ed. Reprint 2001). *Methods of knowledge: perceptual, non-perceptual according to Advaita Vedanta.* Kolkata: Advaita Ashrama.

Sayood, K. (2000). *Introduction to data compression*, Morgan Kaufmann. http://www.csd.uoc.gr/ ~hy438/lectures/Sayood-DataCompression.pdf (visited 2 Dec. 2012

Schutz, A. and Luckmann, T. (1973). *The structure of the life-world.* Evanston, Ill: Northwestern University Press

Shimoda, Masahiro (2010). Some reflections on the history of Buddhist canons in ancient India. In: *Indian philosophy and text science* / Edited by Toishihiro Wada. Delhi: Motilal Banarsidas Publishers P. 33-57

Shuman, Bruce A. (2001). Issues for libraries and information science in the Internet age. Englewood, Colo.. Libraries Unlimited. (Indian ed. 2005).

Sigfried, Tom (2000) The bit and the pendulum: from quantum computing to M theory. The physics of information. New York, Wiley & Sons.

Srinivasan, Radhika (2002). Sacred space: a Journey through the spirit of Asian art. Edited by Edwin Bokert. Foreword by Swami Dayananda Saraswati. Mumbai:Bhratiya Vidya Bhavan.

Sri Ram, Gopal (2007). Information support for the legal profession. New DeLhi: Ess Ess Publications for Sarada Ranganathan Endowment for Library Science, Bangalore

Stanford, R.G. (1977). Experimental psychokinesis: A review from diverse perspectives. In: Handbook of parapsychology, ed. by B.B.Wolman; p.324-381; New York: Van Nostrand Reinhold.

Stehr, Nico and Meja, Volker (2005). Introduction: the development of the sociology of knowledge and science. p.1-27. IN: Stehr, Nico and Meja, Volker. Society & Knowledge: Contemporary Perspectives in the Sociology of Knowledge & Science. Ed.2. New Brunswick, N.J., Transaction Publications.

Szalay, Alexander S. and Gary, Jim (2004). Scientific Data Federation: the world-wide telescope, p.95-108. IN:Foster, Ian and Kesselman, Carl, Eds. (2004). The Grid: blueprint for a new computing infrastructure. Ed. 2. San Francisco, CA, Morgan Kaufman, Indian reprint, Elsevier.

Tapar, Romila (1996). Time as a metaphor of history: Early India. ISBN 139780195637984

ThinkQuest (2007).Time in our lives: The role of time in various cultures

Tobler, C. (1996). Procreation time. Pregnancy and childbirth in EC sex equality law. *Time & Society,* 5(3): 362-383

Toffler, Alvin (1970). Future shock. Bantam books.

Toffler, Alvin (1980). The third wave. Bantam Books.

Trompenaars, Fons sand Hampden-Turner, Charles (1996). Riding the waves of Culture; p.125-126

Valentine, Karen (2005). Schooled for the future? Educational policy and everyday life among urban squarters in Nepal. Greenwich, CT: Information Age publishing Inc.

Valentine, Karen (2008). Mobility, (il)legality and marginalization: Squatting in Kathmandu. IN: Pyakuryal, K. et al. Eds. Social sciences in a multicultural world. Kathmandu, Nepal: Sociological and Anthropological Society of Nepal.

Valentine, Karen (2010). Struggles over school time: Temporal organization of schooling in Nepal and Denmark, p.46-71. IN; Thapan, Meenakshi, Ed. Contested spaces. Citizaenship and belonging in contemporqary times. Hyerabad, India: Orient BlackSwan.

Von Baeyer, Hans C. (2004). Information. Amazon Books

Wada, Toshihiro, Ed. (2010). Indian philosophy and text science. Delhi: Motilal Banarsidass Publishers Private Limited

Webster, Frank (2002). Theories of the information society. Ed. 2. New York, Routledge.

Weirich, Paul (2008). Time preference. p. 369-370. IN; Darity, William A. Jr. Ed. International encyclopedia of the social sciences. 2nd ed.. Farmington Hills MI, Gale, 2008.

Wennerberg, Pinar and Schulz, Klaus (2010). An ontology of socio-cultural time expressions. *Proceedings of the Association for the Advancement of Artificial Intelligence.* Spring 2010, p. 62-66. wwwaaai.org

Wheeler, John. (1989) Information, physics, quantum: The search for links, p.354-368. IN: Hey, Anthony, J.G. Ed. Feynman and Computation.

Wikipedia the free encyclopedia

Reading (Mass.) Perseus Books, 1999. (Reproduced from the Proceedings of the Third International Symposium on the Foundations of Quantum Mechanics, Tokyo, 1989).

Whitehead, Alfred North (1934). Nature and Life. Cambridge, Cambridge University Press.

Whorf, Benjamin (1941). The relation of habitual thought and behaviour to language. (reprinted from Language, Culture, and Personality, Essays in Memory of Edward Sapir, edited by Leslie Spier, p.75-93; Menasha, Wis. Sapir Memorial Publication,1941). (Accessed 9 March 2012) http://www.sloan.stanford.edu./mousesite/Secondary/Whorfframe2.html

Whybrow, Peter C. (2006). American mania: when more is not enough. New York, NY: W.W.Norton.

Wiener, Norbert (1954). The human use of human beings: cybernetics and society. New York, Avon Books, 1954.

Wikipedia; http://www.thefreedictionary.com/time (visited September 2010)

WordNet 2.1

Zahorian, Izzie (2008). Concept of time across cultures. (reprint)

Zerubavel, Eviatar (1981). Hidden rhythms, schedules and calendars in social life. University of California Press.

Ziman, J. M. (1969). Information, communication, knowledge', *Nature*, 224; p.318-324

Zipf, George K. (1949). Human behaviour and the principle of least effort, New York: Hafner.

Index

Note: def = definition; irt = in relation to

A

Acceleration of innovation
    irt Knowledge obsolescence 125
Afghanistan
    irt Expressions of time 61
African cultures
    irt Expression of time 57
Agricultural practice 8, 75, 76, 80
AI 15, 17, 18, 137, 146, 154
Angkor Wat temple
    irt Unity of concepts and terminology 9, 91-4
Arabic cultures *irt* Expression of time 57
Archaeology 11
    irt Time in evidence-based studies 135-6
    Time in historical studies 135-6
Archive 13, 121, 127, 141
    irt Time in evidence-based studies 142-3, 149
    Time in historical studies 142-3, 149
Arctic Circle 58
Art and crafts of Jarkhand 96
Art, Pop 94
Art, Traditional 94-5
Artificial intelligence *see* AI
Arts,
    Tribal 95
    Visual 91
Astrometry, 12, 13, 41, 127, 136, 141, 142
    Theoretical 142
Astronomy
    irt Time in evidence-based studies 141
    Time in historical studies 141

B

Baroque 100, 101
Behavioural aspects 131, 132
Big science,
    Massive Data of 126
Bio-linguistics 154, 155
Biological and psychological effects of time on human development 76
Biological time 76-8
Biostratigraphy 138

Boltzman equation model 10, 122
Bradford law 129
Broader term
see BT
BT 151, 162
Buddhism
irt Expressions of time 63-5
irt Folk arts 95
Business
cultures 9
information 127-8

C

CAI 22
Cambodia 9, 91-2
Carnatic music 97, 99, 100
Categorization of Concepts
irt Semantic relationship in KOT 152
CC 9, 105-6, 116, 152, 156-7
Celestial objects 12, 141-42
Changing mix of information and its social impact 20
Chaos theory 6, 33, 34
Christianity
irt Expression of time 57, 58-64
Chronobiology 76-7
Circadian rhythm
irt Effects of time on human development 76
Classic device *see* CID
Classical
music 9, 94, 96- 100
dance *irt* Hindu religious architecture 92
Indian 9, 92, 101
Western 9
Classics 105
CID 106
Clustering 152
irt Modes of formation of subject 155
Relationship among concepts 155
CMC 51
Colon classification
see CC
Communication, 117
Cultural 38, 117
Electronic 51
Novel medium of 51
Scientific 115, 127
Symbolic 35, 38, 134
Community
knowledge 71
vocabulary 72
Complexity 33
Compressibility and expandability of time 114
Computer-mediated communication
see CMC
Concept
def 16
dictionary 148
of knowledge society 16
terms 136
Concepts-net 152
Consumer-aimed Information
see CAI
Continuity in society 6, 35, 117, 134
Contraceptives 8, 75, 77
Cosmic time 65, 91-2, 94
Cosmological time 63
Cosmology 5, 9, 12-3, 25-6, 34, 92-3, 141-2

Costa Ricans 61
Cultural
  and social time 9, 78
  context 8, 90
  differences 5, 6, 56, 67-8, 164
  frames 1, 164
  impact
    irt System and style 158
  inheritances 117
  landscapes 11, 135
  pre-suppositions 1
Culture, 72, 77
  Material 11, 135-6
  Negative work 30
Culture-dependent time vocabulary 72
Cultures
  of India
    irt Expression of time 61
  without time concept
    irt Time pace 84
Cultures, Indigenous 58
Cybernetics, 25, 155
  Medical 154-5
Cyclic character
  irt Information and entropy 38
Cyclic Time 2,3, 63

D

Dance
  irt Time in arts and humanities 9, 91-2, 96-7, 101, 115-6
Data,
  Compression of 115-6
  Conceptualizing 71
Database, 53, 71, 128, 149
Definition of information 112
Denotative 160-1
Deva years in Indian mythology 65-6
Different forms of temporality 75
Differentiation and equifinality 39
Diffusion 10, 121-5
Digital information 113, 128
Disclosure
  space 118
  time 118
Dissemination and diffusion of information 121
Distillation 152
  irt Modes of formation of subjects 155
Document life parameter 129
Domain 4, 8, 33, 46-7, 53, 74, 114, 132, 150-1, 155, 158, 164-5
  of information science 14, 146
  ontologies, 15, 147
Dviteeya paraardhe
  irt Vedic rituals 66
Dynamic homoeostasis 38

E

Eastern cultures
  irt Expressions of time 57
Eastern Religions/Cultures
  irt Prophesies 70
EBHC 144
EBM 144
Economic Criteria
  irt Cybernetics-Centered Cosmology 27
Educational calendars
  irt Institutional Time 81
Embodied knowledge 5, 9, 16, 112, 150, 164

Encoded information 9, 112
Entropy 38, 39, 117, 121, 134
Environment, Socio-cultural 17, 150
Epidemics 124
Epilogue 164
Epistemological standpoint 8, 14, 74, 145, 146
Epoch 65-7, 69
Europe, 11, 50, 58, 135
    irt Time Studies 141
Evidence-based health care
    see EBHC
Evidence-based medicine
    see EBM
Expandability and compressibility of time 114
Explicit specification 14, 146
Exponential
    and logistic patterns 121
    growth of information 23, 120-1
Extension of
    life expectancy 49
    powerbase 49
    the boundaries of information world 41

F

Fission 152
    irt Modes of formation of subjects 154
Folk
    arts 9, 94-6
    dances of India 102
Folksonomies 15, 147
Fossils 12
    irt Time in historical studies 136-7
Fusion 152
    irt Modes of formation of subjects 154-5

G

Gender-oriented calendars 82
Genealogy
    irt Time studies 139
General systems approach 121
Globalization 29, 46, 83-4
Growth in time: Exponential and logistic patterns
    irt General systems approach 121
Growth of knowledge and information as time-related Properties 120

H

Hadron Collider *see* LHC
Harvest era
    irt Institutional Time 80
Heat conduction model 10, 122
Hebrew culture *irt* Expression of Time 57
Hectic Time
    irt Time Pace in different cultures 84
Hindu religious architecture 92
Hinduism
    irt Expressions of time 64, 69
    irt Prophesies 69
Historical Time 78
Historiography 135
History 3, 11, 118, 134
    irt Time studies 135
Hollingsworth
Homonyms and homographs
    irt Semantic relationships 161
Hormone replacement treatment
    see HRT
HRT 49

Human
- activities 134
- body 8, 38, 46, 74-5, 77, 93
- development 8, 9, 90
- history 3, 11, 41, 135-6
- life 8, 74, 76, 85
- mind 114, 117
- species 5

Hyperbolic equations
- *irt* General systems approach 124

I

ICT 130

Idea
- *def*, 16

Idea acceptance
- *irt* General systems approach 123

Impact
- factor and indices of performance 130
- of complexity on information 46
- of Internet
  - *irt* Internet language 51
- of new media on the public 43
- on human time 48

Increase in government information 46

Increased colonization of time in life course 8, 75

Indian
- culture 64
- languages 53-4, 65, 87, 159, 160
- music 96
- mythology
  - *irt* Expressions of time 65-6
- society 39, 53, 64
- stage
  - *irt* Internet and language 52

Indic culture
- *irt* Time and the arts and humanities 91

Indonesia and Vietnam
- *irt* Expressions of time 61

Information
- and entropy 37
- bound universe 24
- categories
  - *irt* Changing mix of information 21
- communication technologies 134
- diffusion: Heat conduction model 122
- flood 42
- input *irt* negative feedback and coding 38
- networks 28-9
- Science Abstracts
  - *see* ISA
- science 4, 10, 14, 112, 117-8, 136, 146, 164-5
  - *def* 113
- search time
  - *irt* Properties of information 119, 120
- society 5, 24-9, 31, 147
  - *irt* Cybernetics-centered cosmology
- systems 5, 10, 14, 17, 71, 113, 133, 135, 146, 149, 150, 163-4

Information and communication technology
- *see* ICT

Infructuous time
- *irt* Impact on human time 48

Innovation diffusion models 123

Input *irt* Information and Entropy 37

Institution of the Family and Family Time
irt Time / calendars 80
Institutional time / Calendars 79
International Society for Knowledge Organization
see ISKO
Internet
and communication technology-centered society 40
and language 50
languages 52, 53
media and authorship
irt Exponential growth of information 23
technology 51
Interpersonal relationships 7, 69
Iravupanns
irt Time theory 97-8
ISA 10, 112
ISKO 165

J

Jainism
irt Expressions of Time 62-5, 95
Jharkhand
irt Dance 96, 102-3

K

Kannada proverbs 89
Knowledge
and Information 28, 35, 41, 43
by revelation 19
def 17
obsolescence *irt* Acceleration of innovation 125
Organization 1, 4, 14, 145, 150, 165
Tools *see* KOT
society 5, 24, 27-31, 39
transfer 10, 113
Knowledge,
Cultural 4, 8, 14, 74, 146, 164
External manifestation of 9, 112
Intuitive 19
Non-intuitive 19
Knowledge society 16
KOT 1, 4, 150, 164-5

L

Laghu 100
Language
irt Internet 50
expressions, Natural 148-9
Language, Native 53
Lapland 58
Large Hadron Collider
see LHC
Lateral relationships
see LR
Latitude 61, 76
Law 13, 14, 143
irt Time studies 144
Lexicographers 160
LHC 126
Library Science 10, 15, 113, 146
Life
course and time 74
span and actuarial tables 49
Linear time 2-3, 63
Linguistic dispersal of Internet
irt Internet and Language 52
Linguistics, Computational 154-5
Literary Devices 162
irt Semantic relationships 161, 162

Literature 105
Living rhythms *irt* Dance 102
LR 164
Lunar calendar
  irt Indian Mythology 65, 76, 79

M

Machine translation
  see MT
Magnetic resonance imaging
  see MRI
Maithil brahmins
  irt Genealogical Records 141
Manvantara
  irt Measure of time 67
Mathematics
  irt Quantification of information 104
Measuring beats
  irt Dance 102
Medicine
  irt Time studies 13, 143, 157
Metaphorical semantic representation
  irt Semantic relationships 160
Modes of
  formation of subjects 152
  valid knowledge 17
MRI 41
MT 59
Multicultural domain plurality 165
Multilingual Issues 162
Music 9, 21-2, 34, 40, 91-2, 97- 100, 102, 116, 157
Mythology 67, 103

N

Narrower term
  see NT
Natural and cosmic Time 9, 75
Natural time 76, 122
Negative entropy
  irt Information and entropy 38
Netscape 51-2
Networks 28-9, 44, 48-9, 83, 114, 124
New Idea Point
  see NIP
NIP 123
North America
  Hopi tribe
    irt Expressions of Time 61
  Indian tribes
    irt Prophesies 70
Notion and definition of information 112
NT 151, 163, 176

O

Occupational Theory
  irt Cybernetics-Centered Cosmology 28
Old Testament *irt* Expressions of Time 57
Ontological
  approach to representation of time 145
  categories and semantic relationships
    irt Semantic Relationships 162
Ontologies,
  Broad-based culture 148
  Socio-cultural 149
Ontology 14-5, 71-2, 145- 150, 158
  and representation of Time 14, 145
  of SCTE for web applications 147
  of socio-cultural time expressions 148

Other frames for kAlam (time, duration)
irt Expressions of time 67
Output
irt Information and Entropy 37
Overemphasis on Speed 130

P

pakalpanns
irt Time Theory 97
Paleontology 12, 136-7
Parapsychology 103
Pattern of time allocation 81
Perception 17, 18, 104
Phenomena, Cosmic 8, 74-5
Philosophical 8, 14, 62, 65, 67, 74, 90, 145-6, 150
Philosophies 4, 7, 63, 134, 150, 165
Physical Sciences' View 5, 24
Physics 12, 24, 115, 136, 141-2, 157
Planning 50, 75, 83, 85, 130, 153
irt Time pace in different cultures 86
Presupposition 2
Primitive, Arts 94
Private time 42, 82
Production and quantification of information 113
Prophesies about the world 69
Proverbs,
Kannada 89
Sanskrit 89
Tamil 89
Proverbs
as reflectors of wisdom in time-course 86
Psychic abilities (PSI) and time 103
Psychological effects of time
irt Human development 76

R

Radioactive elements 137-8
Ragas 91, 97-9, 156
Reindeer in Sami culture 58, 60, 72-3
Related terms
see RT
Relationships,
Hierarchical 151
Lateral 151-2
Religions 4, 43, 63-5, 148, 149
and cultural viewpoint
irt Cybernetics-centered cosmology 27
Representation of Semantic Relationship in a KOT 151
Requisite information 23, 119
Research communication 10, 130
Research,
Genealogical 139
Scientific 21-2, 118, 125-7
Rhythmic cycles 97
Rhythms 75-6, 98, 100, 102
Rio declaration 44
RT 151, 164, 173, 176

S

Sami
calendar 58, 60
culture 58- 60
time 58
Sankalpam
irt Expressions of time, 67
Sanskrit sayings 88
Science, Big 126
SCTE 15, 147
Sculpture 9, 91-2
Search time 119

Semantic
relationships in culture-specific domains 158
web 15, 71, 146-7
Sensory knowledge 19
Shared conceptualization 14, 146
Slow time
irt Time pace in different cultures 84
Social
construction 77-8
evolution 4, 34, 36, 39, 41
function of communication, 35
justification of large projects
irt Changing mix of Information 22
networks 24, 35, 40, 42, 51, 114, 150
Order, New 39-40
organization 6, 21, 33
processes 8, 90, 74, 107
structures 34-5, 37
system 123-4
time 9
irt Cultural and social time 78
web 15, 147
Societal cohesion 134
and Communication 36
Societal Homeostasis 37, 39
Societies,
Technological 7, 69
Western 82
Society,
Rural 7, 69
Segments of, 29, 63
Socio-cultural
complexity 34
development 34, 46
differences 15, 56, 72, 147, 158
evolution and information 32
evolution of human species 32
time expressions
see SCTE
time 68, 147-9
periods 148
Socio-linguistics 154-5
Sociological
and behavioural aspects 10, 130
implications 123
Socio-technical theory
irt Cybernetics-centered cosmology 28
South American cultures
irt Expressions of time 60-1
Spatial thesis
irt Cybernetics-centered cosmology 28
Statistical patterns in information phenomena 129
Steady state 38
Strata 12, 82, 137
Stratigraphy 138
Structured information 21
Style and system 157
Subject classification 150, 152
schemes 156-7
Subject domains 152
Subsumption relationships 15, 147
Survival in time *irt* Cultural co-operation 36
System and style 157

T

taal 91
taaLa 91

irt Indian Music 99

Tamil

English thesaurus

irt Semantic relationships 158-9

proverbs 89

society, Early 107

Taxonomy 72, 150-1

TDIL 54

Technologies, Communication 26

Technology Development for Indian Languages

See TDIL

Technology growth: Boltzman equation model 122

Temporal shaping of life course 8, 74

Testimony, Verbal 18, 20

Theoretical astronomy 142

Thesaurus,

Bilingual 159, 162

Multilingual 158

Throughput

irt Information and entropy 37

Time

allocation 23, 40, 81

and the arts and humanities 9, 91

as a standard of comparison 39

binding and communication 35

concepts, Socio-cultural 68

for music

irt Indian music: raga classification 97

in

business

cultures 82

data growth 127

collaborative research 10, 126

historical and evidence-based studies 134

information science 9, 112

life course 74

movies

irt Institutional time 81

music

irt Indian music: raga classification 97

Indian culture 64

pace 84

in different cultures 84

period 22, 38, 100, 123, 126, 129, 138

preference 85

or safeguarding one's future interests

irt Time pace in different cultures 85

representation in space 92

zones 98

Time, space and context in text science 106

Timeframe 33, 86

Time-related

English language proverbs 87

Kannada proverbs 89

properties of information 118

Sanskrit sayings 88

Tamil proverbs 89

Time-theory based raga classification 96

Tools, Translation 54-5

U

Unity of concepts 9, 165

and terminology 91

V

Valid knowledge 5, 16-7, 20, 164
  Value approach
  irt Cybernetics-centered cosmology 28
Values,
  Cultural 28, 43
  Social 27-8, 42
Vedic rituals
  irt Expressions of time 66
Vietnam and Indonesia
  irt Expressions of Time 61
Vocabulary words 15, 146

W

Waiting time
  irt Time-related properties of information 118
Wasted time 157
Web applications 148
Weeks *irt* Expression of Time 58
Western
  culture 70, 72
    music 96, 99
      irt Indian Music Classification 100
    philosophy 20
  society and corporate cultures
    irt Expressions of time 57
Wikipedia 11, 15, 43, 55, 57, 99- 101, 135, 139, 141, 143, 146
  irt Internet and language 54
WordNet 2.1 15, 135, 147
Work
  or production calendars
    irt Institutional Time 81
  time 48
Work, Classic 105-6
Worldviews of 21st century societies
  irt Knowledge society 24

Y

Yuga
  irt Indian mythology 65, 69, 93
  Dwapara 66
  Catur 66

Z

Zipf-Bradford distribution 129